Getting Off The Planet

Training Astronauts

Mary Jane Chambers
and
Dr. Randall M. Chambers

To: Dr. Randall M.Chambers, Johnsville, Pa. For: His Contribution to Science
From Mayor Earl C. Lawson, Vincennes, Ind.   August 30th, 1967

We acknowledge the financial support of the Government of Canada through the
Book Publishing Industry Development Program for our publishing activities.
Published by Collector's Guide Publishing Inc., Box 62034,
Burlington, Ontario, Canada, L7R 4K2
Printed and bound in Canada
GETTING OFF THE PLANET
by Mary Jane Chambers & Dr. Randall M. Chambers
ISBN 1-894959-20-5
ISSN 1496-6921
Apogee Books Space Series No. 56
©2005 Apogee Books

Getting Off
The Planet

Training Astronauts

Mary Jane Chambers
and
Dr. Randall M. Chambers

An Apogee Books Publication

DEDICATION

This book is dedicated to our sons, Mark Randall Chambers, and Craig Franklin Chambers, who no doubt remember taking space training aids to "Show and Tell."

It is also dedicated to our daughter-in-law, Reyna Chambers, and to our grandchildren: Randall, Cedric, Sean and Sofia.

AND—To space fans around the world.

The purpose of writing this book is to provide a missing piece of early space history, and to give the scientists, engineers, technicians, designers, medical monitors, flight surgeons, consultants, executives, managers, astronauts, test pilots, and many other supporters, the place in space history which they deserve in the period of 1955-1974.

Credit for the photographs goes to NASA, the U.S. Navy, the U.S. Army and the U.S. Air Force.

About the Title:

The title, *Getting Off The Planet*, is derived from a conversation between the authors. Mrs. Chambers, searching for a catchy phrase for a title, suggested "Getting Off the Ground."

Dr. Chambers, a space pioneer who had not paid much attention to the ground in years, objected.

"The Wright brothers got off the ground 100 years ago!" he said, "space scientists were working to get off the planet!"

Table of Contents

Anyone who glances at this book may wonder about its authorship.

Since Dr. Chambers is a space pioneer and a world-renowned authority on some aspects of space flight training, why is Mrs. Chambers doing most of the writing?

This is a good question.

The answer can be found in the old saying, "Engineers and scientists tend to tell you more about something than you really want to know!"

A case in point is a biologist-friend of ours who spent his entire life studying the chromosome map of the salivary gland of the fruit fly. He was completely unaware that the whole fruit fly—much less his gland—was not of great interest to most of us.

Sending a man to the moon, and returning him safely to Earth in July, 1969, was arguably the most significant scientific achievement of the 20th century.

It was accomplished by scientists from at least two dozen disciplines—from anatomy to zoology— from laboratories of government, industry and universities.

The obstacles to such an undertaking are so enormous that they have seldom been listed anywhere.

Most of us rarely—if ever—think about what it would require to get off the planet. We take gravity, which tightly holds everything to Earth, for granted. We rarely take time to appreciate the fact that gravity is so powerful that nobody has ever fallen off the Earth. Or jumped.

Astronauts in science fiction seem to glide into the great beyond effortlessly. However, the reality is that physical forces such as gravitational stress, atmospheric pressure, and weightlessness all have to be "conquered" before a safe flight can be made.

Beginning in 1958, there were 10 busy years of research—trial, error, testing and rechecking—by scientists of different disciplines to achieve the moon walk.

When Randall became a project director for acceleration research and training of some of the first astronauts in 1958, I found myself with a front row seat to this amazing project.

I was like a theater-goer with season tickets to a series of new, unfamiliar dramas. It took me a while to discover that beneath the footnotes, formulas, charts, graphs and endless data-collecting was a fascinating story.

In connection with his university lecturing, Dr. Chambers is working on two aerospace human factors textbooks which have about 900 pages each. But I decided that space fans, who are not necessarily trained in science, deserve a simpler version of this amazing space adventure.

Dr. Chambers has provided the technical details which I have then tried to translate for us non-scientists.

It has been a real test of our 55-year marriage, almost as difficult as when we tried to wallpaper the dining room years ago.

One might say that this book is itself an experiment to find if authentic science and journalistic human interest can really collaborate successfully.

If not, there's always the textbooks.

—Mary Jane Chambers

I met Randall Chambers in January, 1949, on the beautiful campus of Indiana University where both of us were students. I was a senior journalism major and he was a graduate student in experimental psychology and science.

The journalism classes were conducted on the first floor of Science Hall and the rest of the building was occupied by the psychology classes and labs. The psychologists were considered by many to be the most peculiar group on campus. Some said crazy. The field of experimental psychology was in its heyday and Indiana University seemed to be the Mecca for many far-out experimentalists.

Among them was Dr. Winthrop N. Kellogg, who was raising his young son with a baby ape in order to compare the developmental rates of the two.

Dr. B.F. Skinner, a famous behavioral psychologist who later joined the Harvard faculty, was experimenting with a "baby cage" said to allow an infant to be tended without being touched by humans. He also had pigeons trained to do remarkable things, such as operating levers and tools by pecking on them. He even had trained some of them to launch short-range missiles.

As if this weren't enough, the campus was also the home base of Dr. Alfred Kinsey, who was studying human sexuality. He interviewed male subjects about their sexual habits and prowess and reported the lurid details in scientific journals. It was just a matter of time before the non-scientific press also heard about the studies. Even though the possibilities of exaggeration were inherent in Dr. Kinsey's experiments, the prim and proper world of the late 1940s was shocked.

Dr. Kinsey was a zoologist, not a psychologist, but the public grouped them together.

It was probably a good thing that no one mentioned to this group that scientists would soon be contemplating putting astronauts on the moon.

From the beginning of civilization until the late 1950s, only science fiction and mythology toyed with the idea of sending human astronauts to the moon. If anyone had taken an opinion poll at that time, no one would have predicted that even the experimental psychologists would come up with such an outrageous idea.

In those days, Randy was happily working with Dr. Kellogg doing animal studies designed to understand brain function.

When Mercury candidates and astronauts began arriving at the Navy's Aviation Medical Acceleration Laboratory in 1959, Dr. Randall Chambers (shown here wearing an experimental Mercury pressure suit) was eagerly awaiting with a group of other scientists to start acceleration training, and research in gravity simulators and life support environments.

Dr. Kellogg, somewhat ahead of his time, had published the first book about military pilot training in World War I. Randy described his experiments to me with great enthusiasm. I didn't understand a word he said, but I thought it was wonderful.

At that point in my star-struck life I thought my favorite scientist would discover the cure for some dreaded disease. That was before he met some of the space enthusiasts.

In the summer of 1949, Randy was an air force officer cadet in training at Wright-Patterson Air Force Base in Ohio. There he met Dr. Wernher von Braun, the German scientist who later was called the father of modern rocketry, after having developed the V-2 rocket.

Also in the group was Dr. Sigfried Gerethewold, a pioneer aviation psychologist.

They opened up a whole new insight into the possibilities for manned space exploration.

When Randall and I were married in September, 1949, I was totally unaware that I would have a front seat at the U.S. manned space drama for many years to come.

September 7, 1949 – the first experiment. The groom had a lifelong fascination with science and anything that flies. The bride, who preferred literature to science, was dedicated to keeping both feet on the ground.

We moved to Columbia, Missouri, where he continued his studies at the University of Missouri. He joined another group of experimental psychologists trying to better understand the physical and mental capabilities and limitations of human beings.

Also at the University of Missouri, Randy completed officer training in the air force and received his commission along with his master's degree. President Harry S. Truman presented the air force commissions in person, which inspired the young officers.

Our next move was to Case Western Reserve University in Cleveland, Ohio, where Randy earned his doctor's degree in 1954. Shortly afterward our first son, Mark, was born.

It was at Case-Western Reserve that Randy became fascinated by a newly-developing type of experimental psychology, called industrial psychology. It soon had different labels, such as human engineering, industrial engineering, human factors psychology, or human factors engineering. Later, it included ergonomics, the study of work, the workplace, and the worker.

The military was among the first to utilize this new discipline, calling on the industrial psychologists for help in the design of airplanes, weapons, protective clothing, and training programs to enable work to be done more safely and efficiently.

One of Randy's first projects at Case-Western Reserve was to study the problems of Air Force mechanics, who were charged with the reponsibility of maintaining the B-25, B-29, B-36 and B-50 airplanes. In order to study the problems of the mechanics, he had to learn everything there was to know about the bombers and also fighter aircraft.

He later also studied the problems of air traffic controllers, airport noise, and several projects involving pilot performance and cockpit design. Randy may not have realized it at the time, but he was developing experience and skills that would become essential in preparing astronauts for space travel.

This was during the Korean war and Randy had a mobilization assignment to Wright-Patterson Air Force Base where he had hoped to work with some of the pioneer researchers in the field of human factors. Among these were Dr. Henning von Gierke, a pioneer in bioengineering; Dr. Walter Grether, a psychologist who was director of the human engineering lab, and Dr. Julian Christensen, an early human engineering researcher.

We had just settled down in Bar Harbor, Maine, where Randy was doing basic research at the Jackson Memorial Laboratory, when his military orders arrived. With characteristic unpredictability, his assignment was, not to Ohio, but to Lackland Air Force Base in San Antonio, Texas, another Air Force Research and Development Command Center. He was assigned to the Air Force Personnel Training Research Center of the Research and Development Command, where experimental psychologists continued studying capabilities and limitations of the human body. Among these researchers were Dr. Arthur Melton, Dr. Edwin Fleishman, Dr. Ed Bilideau and Col. Jack Buel.

1955 – With a newly-minted doctor's degree and an Air Force commission, Dr. Chambers moved to San Antonio, Texas with his young family. He was assigned to research projects at Lackland AFB and Randolph Field, where he soon fell under the spell of Dr. Wernher von Braun and other pioneers in space and aviation.

This time the experimental psychologists used pilots and airmen as subjects, instead of mice, rats, dogs and chimps. Most never went back to animal subjects.

The air force laboratories to which Randy was assigned were headquartered at the School of Aviation Medicine at Randolph field. To his surprise, his mentors Dr. von Braun and Dr. Gerethewold were lecturing there. They had been joined by Huburtus Strughold, often called the grandfather of space medicine and Dr. James Guam, a bioastronautics researcher.

As they had done at Wright Patterson Air Force Base, the German scientists often gathered informally with young air force pilots and scientists to discuss the possibilities and challenges of space travel. Their enthusiasm for this was contagious.

In all that has been written about space travel, it has seldom been mentioned that outer space is a dangerous, mysterious, unwelcoming place.

Each generation has had adventuresome types who could not resist the lure of peering down from atop the highest peak, or poking around in the deepest canyon. These explorers were not deterred by fears of polar bears, tropical fevers, man-eating tigers, alligators, jungle rot or other threats.

However, no matter what dangers they confronted, nor how harsh their surroundings, earlier explorers still were nurtured by planet Earth. They could take for granted that the Earth would provide them with air to breathe, drinking water, food, tolerable atmospheric pressure, composition and temperature ranges; and that gravity would hold them to the earth.

Would-be space travelers had none of these life-sustaining amenities—they had to carry life supporting atmosphere with them. The Earth is surrounded by a gaseous mass called the atmosphere. It is comprised of oxygen, nitrogen, carbon dioxide, argon, helium, and hydrogen and traces of other gases. These are life-sustaining elements for all living organisms.

Beyond the atmosphere are the stratosphere and other layers called the troposphere, chemosphere, ionosphere and exosphere. The stratosphere is in reality a void, totally empty of everything, including air. The other layers are similar, but not identical. Each has its own set of astrophysical characteristics and hazards. This complicates getting off the planet because astronauts, en route to outer space, must travel through these five layers and accommodate to each one. Beyond these spheres is infinity, a vast, limitless space with dimensions still unknown today.

In the modern world of thermostatically-controlled central heating and air conditioning, it is hard to imagine adapting to the temperature fluctuations of outer space where temperatures may range from $300°$ C to $-200°$ C below zero.

We are accustomed to 24-hour days. In space travel, the length of a day varies widely, with day-night cycles ranging from 24 hours to 90 minutes.

We earthlings also take gravity for granted. It holds us to earth in the most reliable way. During space travel, there may be gravitational extremes as high as 15G during launch and reentry, to microgravity and complete weightlessness. It is also a landscape which continually moves as the Earth rotates.

Astronomers have been studying the heavens for centuries in an attempt to understand the movements of the planets and other heavenly bodies. To this day, they still cannot issue guarantees that a space vehicle will not somehow—somewhere—encounter an uncharted, unknown danger.

Many people find it difficult to pack their belongings for a week's vacation. Preparing an astronaut to leave the planet involves an enormous effort and attention to detail.

The space traveler has to take with him everything he needs to survive. Basics include breathing air, drinking water, concentrated food and some provision for waste management. He will need a pressure suit which will envelop the body in the atmospheric pressure, temperature ranges and atmospheric composition necessary for survival.

Built into the space suit will also be G-protection, which will enable the astronaut to survive the gravitational stresses during launch, when the spacecraft breaks free of Earth's gravity, and then again when it reenters like a meteor. Included in the life support system will be biomedical monitoring devices attached to the body. These devices, which monitor heart rate and blood pressure, were particularly useful during the early flights when it was unknown how humans would react to the space environment.

Many people have been under the impression that aviation and space were closely related and practically interchangeable. This is not entirely correct.

Pilots, especially those who flew high altitude planes, had protective clothing developed especially for them. But space conditions, especially the gravitational stresses and wild temperature fluctuations, required special protective clothing made of scientifically-designed materials for space pilots.

Since the life support system literally provided all of the life-sustaining necessities, it had to be practically failsafe or repairable in flight. It also had to be sturdy enough to survive the rough treatment and harsh environment into which it was thrust.

Another area which had to be developed was communications, both with ground control and the monitoring stations. The state of communications in the 1950s and 60s was in the infancy stage. Computers, which are taken for granted in the 21st century, were big, clumsy, slow and unreliable during this period.

However, astronauts could not be launched into space and abandoned in its vastness by ground control. Communications were an absolute necessity. The old-fashioned compass soon was to be augmented by a vast array of navigational developments. And—reversing the growth order of most things— computers began growing smaller and smaller.

It soon became clear that the early astronauts would have to have outstanding qualities of character, personality, intelligence and human-performance capabilities. So the criteria for astronaut selection would have to be established.

It would also be necessary to understand what should be included in an astronaut training program and then to set it up. Obviously, the ability to survive and work in environments that no human had ever experienced before was not easily measured. Nor the ability to adapt to constantly changing conditions encountered in maneuvering in outer space. This may sound like "the right stuff" but actually the right stuff had not yet been determined.

The scientists had to first decide what kind of person could best do this hazardous and exciting job. It was the behavioral scientists who established the criteria for astronaut selection. These space pioneers had to figure out what characteristics would make up the "right stuff."

The third part of the challenge was the spacecraft itself and the rocket which would launch it into space. This included an enormous array of towers, trucks, aircraft, monitoring devices, flight controls, controllers, and landing preparations and recovery. U.S. flights up to 1972 all landed in the ocean. This required an armada of ships, aircraft carriers, boats, helicopters and deep sea divers. At times, this aspect of the manned space flight project—the hardware—seemed to be the most problematic of all.

Designs for cockpits and cabins for spacecraft had not been tested. Cockpits and cabins designed for airplanes were modified and upgraded with each new generation of plane. But everything about a spacecraft was so different from an airplane that there was very little that could be adapted.

For the earliest flights, the astronaut would travel into space lying on his back, with his knees pulled up in a bent-safety-pin position. This was necessary because the body would be able to tolerate more gravitational stress from this position than if the pilot were seated upright.

This meant that the entire cockpit would have to be designed so that the astronaut could operate the controls from that unusual position. There was no doubt it would be a cabin such as had never been seen before. Airplane technology was inadequate for a vehicle which would have to withstand the enormous stresses of travelling into space and return. Particularly troublesome was the question of material for a heat shield which was a special protective covering for the spacecraft.

It had to be light enough for space travel and yet able to withstand the high temperatures of breaking through the Earth's atmosphere during reentry.

A rocket which could be trusted with a manned space capsule had not yet been developed. The science of rocketry, having been developed thus far for firecrackers and weapons, was thought to have almost insurmountable problems in propelling a manned spacecraft into space.

Indeed, the rockets under development in the late 1950s blew up on the launch pad with distressing frequency. Even in 1960 when President John F. Kennedy announced in his inaugural address that going to the moon would be the nation's goal, the failure rate of rockets was over 61%.

Superceding the concern for the patient, meticulous solving of all of these known obstacles was the chilling possibility that some unknown danger—unseen, uncharted, unexpected—was lurking in that mysterious space environment.

In the 1950s—as it had been for centuries—the moon was a distant planet, shrouded in mystery and romance, the subject of love songs and poets. Anyone thinking that sending astronauts to the moon was a serious project would have been thought to be demented or misinformed.

Both lay persons and poets could have been forgiven for feeling that the moon would remain forever out of man's reach. However, to the small group of pilots and aviation psychologists who met with Wernher von Braun and the other German scientists in the air force seminars of the 1950s, the obstacles were not insurmountable, merely challenging. They found the idea of manned space exploration irresistible because the obstacles were so overwhelming.

Those young air force pilots and scientists, inspired by Dr. von Braun and the enormity of such an undertaking, spent many hours discussing the problems—and possible solutions.

When Dr. Chambers' tour of duty ended in August, 1957, he had a number of job offers. With two small sons and a wife who was tired of moving around, Randy accepted a position as an associate professor of experimental psychology at Rutgers University in New Brunswick, New Jersey. (He did, however, remain a captain in the Air Force reserves.)

We moved our family to New Jersey in time for the 1957 fall semester and settled in, prepared to enjoy a quiet, academic environment. It was not to be.

In November of 1957, the news hit the American scientific community with the force of a meteorite striking the earth: the Russians had successfully launched their first Sputnik! The American scientists had been discussing the possibility of exploring space—casually and theoretically—but their Russian counterparts had accomplished it.

Moreover, their technology, in a closed society, was a closely-guarded secret. The Americans were puzzled over every detail—outdone in many aspects of the project. They were like a group of cyclists who had just been run off the road by a fancy new sports car.

How did they do it?

This question was raised by scientists of many disciplines in institutes, laboratories and classrooms all over the country.

The Sputnik launchings were humiliating to the American scientific community. For years they had enjoyed the reputation of being the best—good old American "know-how" was revered throughout the world.

The Germans also prided themselves on their high-tech achievements.

And now both groups were stunned by the spectacular achievements of the Russians. To make matters worse, the details of how orbiting Sputnik had been achieved were a closely-guarded secret in a closed, rather secretive society.

Sputnik I, followed by Sputnik II, were the topics of many meetings in various parts of the American scientific community. And then they began making plans to do intensive manned space exploration of their own. They knew what the problems were, for they had been discussing space travel for several years. Where to begin, however, had yet to be determined.

Dwight David Eisenhower—one of the most famous generals in history—was then president of the United States. He and his staff viewed the launching of Sputnik as a significant military threat.

He often said that the nation who ruled outer space, ruled the world.

There were a few American scientists who were already at work on selected aspects of solving the space puzzle.

Air Force Col. David Simon, a physician, was making a name for himself with his experiments with high-altitude balloon flights. He used himself as a subject and went aloft for long periods suspended in a balloon gondola.

This work was designed to study the effects of a very high altitude environment on the human body and to help determine if humans could indeed survive in such surroundings for a sustained period. Many people were concerned that the balloons used in the experiment were so paper-thin that they might burst, with fatal consequences.

Some thought that Col. Simon was very brave and others thought he was fool-hardy. Navy Capt. Norman L. Barr conducted similar high-altitude balloon experiments for the Navy.

Another space researcher of the time was Air Force Col. John Paul Stapp, who became famous in research circles for his work with an experimental sled. He attempted to study the effects of high G acceleration forces—and rapid, sudden deceleration—on a ground-level track.

He rode the sled at high speeds in the New Mexico desert and studied the effects of sudden deceleration when he brought the sled to an abrupt stop. Most of the time he used himself as the subject of these experiments, making him one of the early space pioneers. Another pioneer in space research was Dr. Randall M. Chambers who was one of the few aviation research psychologists of his era. His specialty was so rare that the Air Force contracted him to set up a lab at Rutgers and then shipped a lot of research equipment to him when his tour of active duty was over. His contributions were on the effects of flight on pilot skills, physiological tolerance, and performance capabilities. He became one of the world's authorities in human performance in unusual environments and also space and aerospace flight simulation.

There were other small groups of scientists in various parts of the country who were doing "space" research, mostly out of fervent interest in the subject. Most of these were military projects. But manned space research in the United States at that time was more theoretical than actual.

The Russians launched Laika, the dog, in Sputnik II, sending chills of apprehension through the American scientific community and the political establishment. The thinking was this: from a military standpoint the progression for weapons had always been first the vehicle, then the passenger, and then the bomb. Not only did the American scientists agonize over the scientific edge of the Russians, they were also concerned about the military threat.

The United States Air Force had begun its own military space project several years earlier. It was called "Cis-Lunar" meaning "to orbit the moon." Randy had worked on the Cis-Lunar project when he was on active duty with the Air Force. The Navy also had announced a military space program.

At Rutgers Randy had an experimental psychology laboratory which contained the equipment from the Skills Component laboratory of the Air Force Research and Development Command. He also had a group of excellent graduate students. At that point he would not have dared admit that the quiet, peaceful academic life was just that—too quiet and peaceful. It had always been apparent that he was restless in the campus environment and that he interacted with the faculty like a lion cub amid a litter of kittens.

Sputnik—especially Sputnik II with the dog aboard—settled the matter. Rutgers gave him tenure and even assigned him his own parking space. But the lure of space research beckoned irresistibly and he accepted the position of head of the Human Engineering Division of the Aviation Medical Acceleration Laboratory of the U.S. Naval Air Development Center in Johnsville, Pennsylvania.

Some of his Rutgers work was also incorporated into his new research. With our sons, Mark, age 4, and Craig, age 1, we moved to Hatboro, Pennsylvania, in the summer of 1958. Our brief stint at the serene academic life had come to an abrupt end.

At the Naval Air Development Center they were rapidly gearing up to do space research. At about that time a new government agency—called NASA—was created, replacing NACA, the National Aeronautics Committee Advisory group, which had begun in the early 1900s.

We moved into a two-story, red brick colonial-style house in a pleasant residential neighborhood. We tried to fit in and pretend that our lives were as prosaic as that of the teachers, bankers, and businessmen around us. When people asked me what my husband did for a living, I tried to mumble so they wouldn't understand that I was saying, "He's doing research to send astronauts into space and to explore the moon."

Hatboro, Pennsylvania, is a suburb of Philadelphia, a center of rich historic traditions. In the 1950s, like most parts of the country, traditional life conducted wholly on this planet, was the prevailing attitude. Sputnik had not disturbed—nor inspired—the general population as it had the scientific community. It soon became impossible to pretend that ours was an ordinary household, with a 9-to-5 breadwinner.

The Mercury Project was on the drawing boards and the acceleration research was centered in the Pennsylvania facility. Parts of the research effort were parceled out to scientists from NASA Langley Research Center Space Task Group, industry, university laboratories, and other units of the government. Soon the small town was filled with visiting scientists, test pilots, various types of engineers, and other researchers. Later on they were joined by astronaut candidates. This entire group stood apart almost as much as if they were space aliens from another planet. Our household soon became the unofficial headquarters of space science, a monument to individualism in a neighborhood of dedicated conformists.

There is no place on Earth that resembles the space environment, so in order to train astronauts and prepare them to survive launch and reentry and to work in the weightless environment, the training was to be performed on simulators and research training devices. The major simulator among them was the giant human centrifuge at the U.S. Navy's Aviation Medical Acceleration Laboratory at the Johnsville center.

I had never heard of a human centrifuge until we moved to Pennsylvania, but it became the centerpiece of acceleration and gravitational research—and of our lives—for the next 10 years. Most people became acquainted with space simulators from the movie "Apollo 13" when they were used to rescue the damaged vehicle. However, simulators were vital parts of space research from the very beginning.

The human centrifuge had been used for years to study the problems of fighter pilots who climbed high into the stratosphere. Studies on the centrifuge enabled researchers to determine the requirements for protective clothing, life-support equipment and training which would allow them to function in an oxygen-deprived environment. Now it was being used to answer crucial questions which had to be answered before the Mercury program could go forward. Foremost among these was whether the human body could withstand the gravitational stresses involved in launch and reentry—reaching

17,000 miles per hour while their spacecraft was subjected to wild temperature fluctuations between several thousand degrees heat and several hundred degrees below zero.

Unless the human body could adapt to these conditions with special materials, protective clothing, a reliable life-support system and training, manned space flight would be deemed impossible. The entire concept of manned space travel and human survivability rested upon the findings of these centrifuge studies.

If these questions were answered favorably, the next round of study would be fine tuning. The capabilities and limitations of the human pilot to withstand and function under many different conditions of gravitational stress, including angular and linear accelerations, vibrations, rotations, and oscillations, had to be established to specify the design parameters for the manned space vehicles before they could be designed. Every item of space equipment had to be researched before it could be designed.

In most cases the materials from which everything was to be made also had to be researched, designed, or invented. Then tested and evaluated again.

The human centrifuge is housed in a big, round building all its own. From the outside, it looks like a giant layer cake with gray icing. The centrifuge itself resembles a bizarre carnival ride, although it is much more complex. A 50-foot arm of bridge-girder steel moves in a circle like the hand of a giant clock. One end of the arm is attached to a large round cement platform in the center of the circle. A rotating gondola shaped like a space capsule is attached to the other end of the arm. The trainee rides in the gondola which whips around the outside of the circle, narrowly clearing the walls. An observer platform on the side of the chamber and another platform in the ceiling enables technical monitoring of the subject.

The giant human centrifuge was housed in this unusual round building, the Aviation Medical Acceleration Laboratory, at the U.S. Naval Air Development Center, Johnsville, Warminster, Pennsylvania. It could be used as a total mission simulator for research and training in aviation and space. During simulation of space flight missions, it could provide additional research and training support in many specialities: aerospace medicine, human factors, engineering psychology and human performance, physiology, biophysics, biomedical engineering and life support, computer aeronautics and engineering technology.

The Aviation Medical Acceleration Laboratory's human centrifuge. It superimposed and combined stresses of linear and angular accelerations and gravitational accelerations. Thousands of hours were spent in this human centrifuge by scientists, pilots, astronauts and engineers studying and training for the problems imposed by the acceleration stresses of launch, reentry and gravity environments.

Each platform is equipped with a manual control switch which could be used to override the regular electronic controls. The subject climbed into the gondola from the side loading platform and exited the same way.

The interior of the centrifuge room is round, with cement walls, floor, and ceiling. Because of its size, it has the cavernous feeling of a coliseum. It also has the gray roughness of a prison, the intricate wiring of a telephone relay station, the crisp, white-coated efficiency of a hospital, and the busy excitement of a genuine launch pad.

In the fall of 1958, the first centrifuge simulation was being planned to train the Mercury astronaut candidates. There were many variables: the spacecraft which was being simulated had not yet been constructed; human reactions to acceleration stress were unknown, and nobody had ever flown a space mission.

This was a happy moment for Dr. Randall Chambers. He had done experiments on animals, men, machines and airplanes. Now he was faced with the king of machines, the knottiest kinds of problems, and a group of well-trained, dedicated pilots as his subjects. It was his job to help combine these things into a research and training program which would prepare astronauts to travel into outer space and back again safely.

We had our first space argument the week he began his new job. "What are you going to be doing?" I inquired innocently. "Well, I'm Project Officer for the acceleration training of the Mercury astronauts."

"That means I'm going to help prepare the astronauts for space travel," he added.

"Is it dangerous?" I asked.

"Well . . .," he hesitated, "not much has been done in this field. We'll know more after I ride the centrifuge." "After you ride it?" I exclaimed, "I thought it was the astronaut candidates who were being trained."

"I can't train them to do something I don't know how to do myself. Besides, I don't want anything to happen to them while I'm in charge."

I could only protest weakly, "Lots of obstetricians have never personally had babies, but they seem to be able to do their work all right."

My argument, as logical as it seemed, fell on deaf ears. Randy was among the first to ride the centrifuge, along with a small group of other pioneers: psychologists, biophysicists, engineers, flight surgeons and assorted other volunteers.

They were joined in 1958 and early in 1959 by a group of test pilots from NASA, the Marine Corps, the Navy and Air Force, who participated in centrifuge training and evaluation studies. Along with preparations for manned space travel were ongoing studies of G-suits to protect pilots from gray-out, black-out and unconsciousness at high G levels.

This work served as a basis for acceleration training and simulation programs for astronauts and pilots during the next 15 years.

In these early studies, test pilots, scientists and astronaut candidates spent hundreds of hours testing and practicing many varieties of flight maneuvers on the giant machine. This gave the astronaut candidates the opportunity to practice various launch, reentry and other maneuvers under varying conditions of gravitational stress.

The interior of the centrifuge gondola, where the man rides, looks very much like the inside of a space capsule, and it is fitted out with a control panel which is a duplicate of the real thing. During the Mercury simulation, the gondola had a Mercury control panel. Other panel configurations followed for each new project.

During a centrifuge "run," the arm whirls around at a high rate of speed and the gondola can be made to pitch, roll, and yaw. Vacuum pumps reduce the atmospheric pressure to add further authenticity to the space rehearsal.

High on one side of the centrifuge room is a glass booth which looks like the compartment for the press at a football stadium. This is where the medical monitor sits with his instruments for measuring the bodily reactions of the man in the gondola. When the centrifuge turns, the trainee experiences the same acceleration stress which an astronaut encounters when he is launched or when he reenters the atmosphere.

The force of gravity which holds everything to the earth, does not give them up easily. The space vehicle breaks the bond of Earth at speeds of up to 17,000 miles an hour. During the period of launch and reentry, the weight of gravity presses against the astronaut's body with a force many times his own weight. This gravitational stress—known simply as "G"—abuses the body. It causes pain in the chest, arms and legs; distorts vision and speech; makes breathing difficult, and breaks little blood vessels in the eyes and the skin. It also makes the heart work harder.

By "flying" the centrifuge, the researchers were able to learn how to resist the effects of acceleration stress through the use of proper breathing and positioning techniques. After they had practiced this, they learned how to concentrate on performing their tasks during exposure to stress. My space education came gradually and I didn't know all of the facts of life about acceleration stress when Randy and his fellow scientist-space enthusiasts spearheaded the centrifuge studies.

This was probably just as well. They tried riding the centrifuge in a variety of positions. They rode backwards and frontwards, spinning and turning, lying down and sitting up. They rode blindfolded and they rode while under the protective influence of experimental drugs. They even rode following prolonged immersion in water. They rode in different positions in order to compare acceleration stresses chest to back, back to chest, side to side, head to foot or foot to head. They tried out various experimental biomedical sensors to measure the pilot's blood pressure, respiration, muscle tension, temperature, brain waves and other reactions, as well as his performance under stress. They also developed and experimented with devices to protect the body from acceleration stress. These included contour couches, G-suits, thermal suits, pressure suits, visors, goggles, helmets, gloves, seat belts and harnesses, biomed sensors, aircraft seats, and air and water suits.

Most of the experimental riders did not admit to being frightened, although hair-raising reports of centrifuge rides were not uncommon. Some of the subjects blacked out during the ride and had to rest for several hours. Some became disoriented and were unable to walk unassisted. Some had unusual perceptions and sensations. Sometimes the damaged blood vessels in the eyes and skin were painful for several days. Some suffered from nausea and there were many jokes about how many times centrifuge riders lost their lunch.

I finally grew accustomed to having a husband who came home from work covered with scratches and bruises. And I even got to the place where I could keep from wincing when Randy and his friends talked about experiments they called by names such as "Eyeballs in, Eyeballs Out . . ."

As the centrifuge work progressed, Randy and his co-workers tested many types of pressure suits. One of these, fitted especially for Randy, was the project Mercury space suit, which was later worn by the first astronauts. When I first saw the space suit, it looked very strange—like something out of Buck Rogers. But after Randy had brought it home several times it began to look as ordinary to me as a pair of overalls would look to a mechanic's wife.

I should have realized that the neighbors would need a period of adjustment to the space age. However, I didn't think of this until one Friday afternoon when the paper boy came to collect his money. A space suit, complete with helmet and gloves, was laid out on a chair in the living room, giving the impression that we had a visitor from outer space.

"Sixty-five cents, please," the boy said, with a guarded glance in the direction of the space gear.

When I handed him his money he took it nervously and, then, with another glance at the space suit he ran out the door and down the street. He couldn't wait to tell the neighbors that he suspected that a flying saucer had just landed nearby and that a Martian was visiting the Chambers' house. Later that afternoon the boy returned with a group of his friends and they examined our yard carefully, as though looking for some clues to our celestial visitor.

As I saw more and more of the centrifuge and the space gear, I began to see less and less of Randy. Not only was he serving as one of the chief subjects in the centrifuge experiments, but the job of project director involved being coordinator, trouble-shooter, and peacemaker as well.

In his laboratory, Randy encountered as many crises in a day as the diplomatic service.

He worked with several Washington bureaus, and with the Navy, Air Force, NASA, other government laboratories, aerospace industries, and universities. He was continually interrupted by visitors, official and unofficial, foreign and domestic; and by telephone calls, temperamental technicians, budgetary problems and personnel conferences.

The crowd of people connected with the centrifuge projects seemed to be continually expanding: the centrifuge crew, the medical monitors, the flight surgeons and the computer experts kept increasing in numbers and enthusiasm. Then came the engineers from the industries which had manufactured the capsule and other equipment; various specialists from the Army, Navy, Air Force, and NASA, and, from time to time, admirals, presidential advisors and congressmen. And as the astronauts became better known, their arrival was heralded like that of a company of movie or rock stars. Everybody and their distant cousins wanted autographed pictures. Reporters wanted interviews. Television and radio commentators wanted coverage. And before long there was another kind of space problem: standing room only in the centrifuge building.

The centrifuge was driven by a computer system, located some distance away from it, operated by electronic engineers and mathematicians. Each of these specialists performed his specific job at a monitoring station. Some of these stations were located in different parts of the building, and others were farther away and were connected by Dataphone cables. There were computer monitors, medical monitors, powerhouse monitors, performance monitors, pressure suit monitors, temperature monitors, instrumentation monitors, recording stations, a data-taking station—all coordinated by the project director.

Despite all these monitors who slaved over it like physicians to royalty, the centrifuge occasionally got out of hand. There were literally hundreds of things which could go wrong with it—and some weeks it was out of order more than it was working. An ordinary machine would have broken down and stopped running. But the centrifuge was an extraordinary machine and when it broke down it did so by running faster. Heady with the energy of its 1.5 million horsepower, it would ignore the electronic controls, kick over the traces, and do some experimenting on its own. One of its favorite tricks was to run at a higher G than the researchers had called for.

On one of its more piquant days, the centrifuge took an innocent victim who was practicing for launch and stood him on his head. Then, as if to compound the joke, it stopped dead still, leaving the poor man suspended upside down. Fortunately, the astronaut candidate was a veteran test pilot, who remained calm about his undignified predicament. When the centrifuge ran away with itself, the research projects came to a complete standstill until the trouble was located, diagnosed and remedied. This sometimes took days—or even weeks. Much has been written about stressful jobs, but it would have been hard to find a workplace with more commotion than the centrifuge building in those days.

One day I went over to the laboratory when they were getting ready to do a simulation experiment. Someone from NASA had ordered some special equipment installed on the centrifuge. But local technicians were threatening to quit if this was done. Other technicians were working frantically over a myriad of wires, trying to find the source of some other trouble. I asked one of Randy's co-workers how things were going.

"The most terrible things keep happening," he said, shaking his head dazedly.

But Randy looked up, smiled, and said, "Everything is going to work out just fine."

This ability to be calm and objective is a prerequisite of space work. The astronauts have this objectivity and so do most of the scientists working behind the scenes. The centrifuge pioneers worked for about nine months before the first astronauts came to begin their training. During this time, they had learned many things about the physiology of acceleration stress and the ability of men to perform under high G exposure. Through trial and error, they had been able to select the best positions for the pilot to be launched, the most convenient placement of controls, and the most effective suit and biomedical measuring devices. They also had turned over to the design engineers a lot of information needed to design the space capsule.

The Mercury astronauts arrived for their first encounter with the centrifuge in the spring of 1959, and soon the names of John Glenn, Alan Shepard, Gus Grissom, Scott Carpenter, Gordon Cooper, Wally Schirra and D.K. Slayton were familiar to all of us. On their first visit, the seven astronauts "flew" 153 static runs and 147 dynamic runs on

Astronaut Alan Shepard during early dynamic flight familiarization and training at the Aviation Medical Acceleration Laboratory in 1960. In this picture, autographed by Alan in 1960, he expressed sincere appreciation to Randy Chambers for his help in the training program. This was the first of eight AMAL–NASA Centrifuge Simulation and Training Programs for Mercury flight dynamics.

the centrifuge, wearing newly-designed pressure suits and flight suits, and using form-fitted contour couches for G protection. This period was devoted mostly to practicing launch and reentry acceleration, and evaluating some of the cockpit controls, the location of switches, instrument displays, restraining equipment and flight procedures.

The various components of the man-machine system that went into the Mercury space flights were being built in a number of places. The Mercury capsule, the pressure suit, the flight control stick, the astronauts' couches, the life support system, the computer equipment, and the biomedical and performance monitoring equipment all had different manufacturers in widespread geographical areas. Each manufacturer tested his equipment extensively and it would all be brought together for a period of pre-flight testing at Cape Kennedy.

A number of scientists and engineers thought that this was sufficient. However, a small handful of pioneers, including Randy and John Glenn, felt that the astronaut should have an earlier introduction to the equipment he was expected to use in space—that once he was in the final phases of preparation it would be too late to discover that modification on some piece of equipment was needed.

The centrifuge was then the only facility in the country where most of these components could be tested on the astronauts at one time. In this realistic space environment, atmospheric pressure, acceleration forces, temperature, breathing air conditions, noise, lighting, instrument displays and flight control equipment could be tested in realistic detail. Modifications could then be made on the spot.

John Glenn, a highly-experienced test pilot, was also an avid centrifuge rider. Some of the other astronauts complained, only half-jokingly, that the centrifuge training was unnecessarily rough or too lengthy or too demanding; but Glenn practiced riding "the wheel" as he called it, hour after hour. It turned out to have been a very fortunate decision.

Glenn did much of the early testing and evaluating of centrifuge runs. One of his chief contributions to the early space program was his insistence that the equipment be fitted around the astronaut early in the flight program—and that the astronaut should participate in the early design and testing of his equipment. This turned out to be such a successful arrangement, that this type of astronaut-equipment-environmental-stress testing was used in subsequent centrifuge programs for Gemini and Apollo.

The Mercury astronauts were not going to have much control over their capsule—riding in it was almost like being shot out of a cannon. But in case the automatic system failed, the astronaut would have to take over. So one of the first performance assessments made on the centrifuge was the determination that a pilot could satisfactorily perform a manual entry into orbit—and subsequently a

Project Mercury astronaut John H. Glenn, Jr., Lieutenant Commander, USMC, of New Concord, Ohio, wearing Project Mercury full-pressure suit, shown prior to being subjected to Mercury-type acceleration and low pressure profiles in the gondola of the human centrifuge.

manual reentry to Earth. The performance of the reentry manually, instead of automatically, was called the "fly-by-wire" technique. This technique was a new concept at the time of Mercury and was extremely difficult to master. All of the operations which had been performed by both hands and both feet of a pilot in an airplane were now to be done by the right hand alone. If a pilot applied too much force to the flight control stick, or if he applied it at the wrong time or at the wrong rate, the spacecraft would go into an incorrect entry attitude, spin, tumble or possibly into severe oscillations. John Glenn, who perfected this technique, was also the first one to use it. (It also helped to save Apollo 13 many years later.)

On his first flight, Glenn's spacccraft developed a malfunction and he took control of it and used the manual reentry technique, which he had practiced on the centrifuge, to bring it down safely.

In his official flight report he later wrote, "I could note no difference between my feelings of deceleration on this flight and my training sessions in the centrifuge."

After a year of intensive, meticulous research on the centrifuge, the crucial questions had been answered: space travel for human beings was far from easy. Indeed, the variables which had to be addressed by scientific study were among the most daunting in the annals of science. But the basic centrifuge studies had determined that the project of sending human beings into outer space was not impossible. Or, to put it another way, challenging but surmountable.

The story of how the U.S. manned space program began cannot be told completely without somehow showing some of the countless projects which were conducted to answer questions, both large and small. However, if we attempt to place these projects in the context of the story, it would be like taking detours on the highway.

So, in Chapters 3 and again in 10 we will stop the narration and present an assortment of pictures of research projects which give depth and scope to the complicated history of how this wondrous endeavor was accomplished in just 10 years.

It is impossible to estimate the number of research projects which were conducted in preparation for sending humans into space. The thoroughness by countless scientists representing assorted disciplines from university, industrial and government laboratories has seldom been equaled. Getting off the planet required years of intensive research which included data collecting, testing and more testing. The plan was to test the whole mission in real time, simulating every aspect of the flight including the astronaut "rehearsing" the operation, complete with flight suit, biomedical monitoring and equipment.

Included here are glimpses of research projects which are not easy to categorize. Some were used in actual space flight; others were set aside for future consideration.

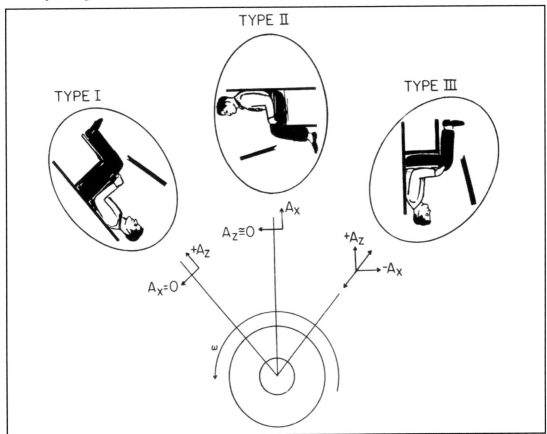

One of the first questions to be answered was this: which position would enable a pilot to withstand the most acceleration stress? Acceleration tolerance limits for high amplitude and duration were tested in pilots seated in the gondola of the human centrifuge for each seated position. The Type II position provided more protection than either Type I or III. At high acceleration amplitudes and for longer endurance times, pilots could withstand the high G-forces and perform better if they were placed in a transverse position with respect to the resultant acceleration vector.

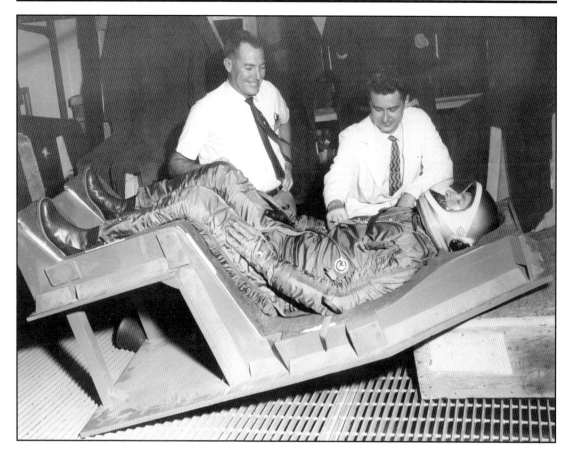

(Above) An astronaut candidate selected the Type II position after testing them all. This position for transverse G became the choice of the pilots. In the early Mercury program in 1960, the astronauts were tested in their early form-fitted couch designs for physiological comfort, circulatory and possible orthostatic hypotension effects, and life support. Brent Creer (NASA Ames Research Center) and Dr. Randall Chambers (Aviation Medical Acceleration Laboratory) discuss and evaluate physiological and psychological aspects of comfort, work-rest cycles and positions, human factors and safety design, for an astronaut interfaced within a pressure suit, helmet, gloves and boots, in his form-fitted contour couch.

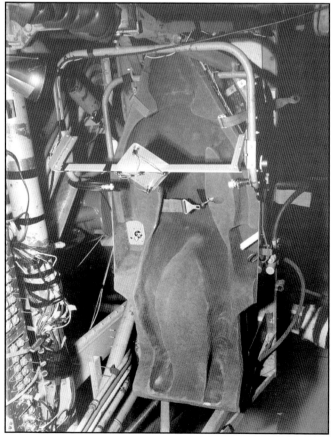

(Right) In the early tests, it was discovered that the subject could tolerate high-G stress longer while encased in a contour couch of the type shown here. However, the couch was too long to fit in a Mercury capsule – so it was back to the drawing board.

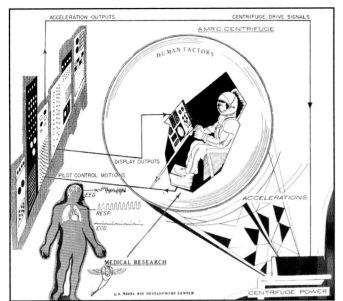

The human centrifuge was the centerpiece of the early space research to determine if humans could actually survive a trip into space and return safely. For the scientifically-inclined, here is a description of the role played by the centrifuge and how it works. Pictured is the coordinate converter system for the human centrifuge, used as a space flight simulator. Acceleration outputs are converted to centrifuge drive signals to produce realistic flight accelerations in the gondola at the end of the centrifuge arm. The pilot's controls and displays in the cockpit are included in the acceleration outputs. These centrifuge simulations of flight provide medical, psychological, human factors and engineering data.

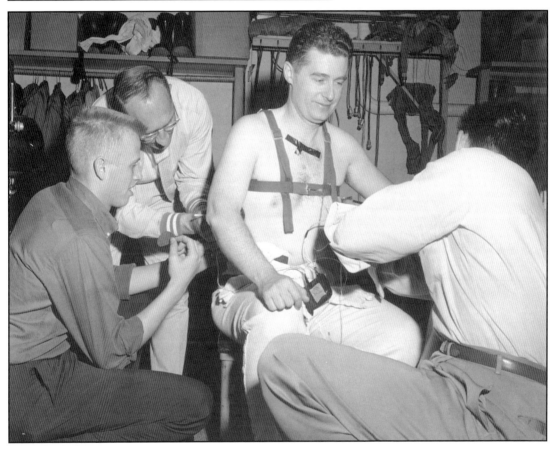

Dr. Chambers prepares for Mercury space flight simulations in the gondola of the AMAL Human Centrifuge. Circa 1960, he studied the performance and protective aspects of the Mercury pressure suit, helmet, gloves, couch, life supports, instrument panel, and controls during simulations of launch, staging, flight events, orbit, and reentry. High-G aborts, escape maneuvers, emergency and monitoring operations were also studied.

NASA Test Pilot Robert Champine in the AMAL human centrifuge, being tested in his form-fitted contour couch, secured by restraint straps for his chest, shoulders, arms, waist and legs. His head is secured by a helmet fitted into the couch, secured by two large bolts. Face, mike and chin restraints are also applied. Dr. Randall Chambers was the project officer, directing the centrifuge test program for AMAL and NASA Langley.

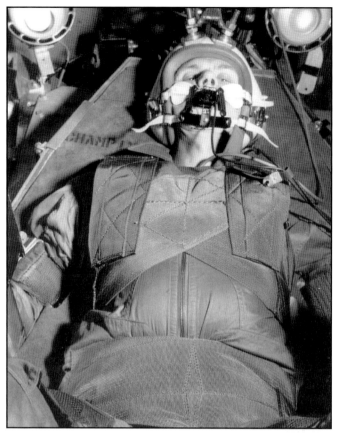

Mercury Astronaut John H. Glenn, Lieutenant Colonel, USMC, practiced many Mercury flight control and life support maneuvers in the human centrifuge during acceleration training and refresher training in preparation for his Mercury Friendship 7 flight, February 20, 1962.

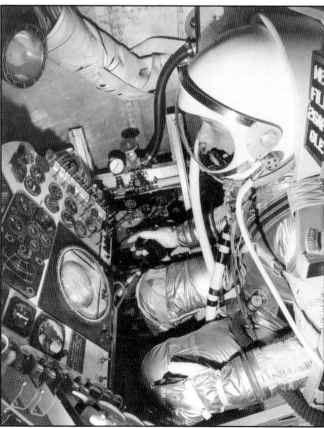

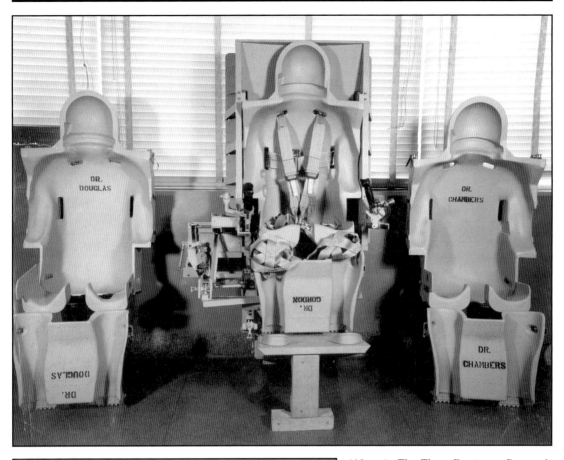

(Above) The Three Doctors – Research scientists not only planned projects, they often tried out the experimental requirements before asking astronauts to perform them.   Pictured here are form-fitted Mercury contour couches belonging to Drs. Douglas, Gordon and Chambers.

(Left) Astronaut Wally Schirra is pictured as he prepares for a centrifuge run in April, 1960, in preparation for a Mercury series of acceleration flight profiles. His helmet features a microphone which allows him to communicate, and a thermistor which gave medical personnel a way to measure his respiration and body temperature. Another feature is a neck ring which allowed him to turn his head.

This picture depicts an underwater tilt table experiment simulating weightlessness at different angular positions. The purpose of these studies was to study the subject's orientation abilities while suspended at different angles at the neutral buoyancy point. Humans are accustomed to living in an environment in which gravity provides location cues. In preparation for space travel, many studies were done to determine if subjects could learn which way was up without the help of perceived gravity.

How large should the Mercury capsule be? And what shape? There were various designs, offering a space cabin of several different concepts. This pictures one of several prototypes which attempted to discover the minimum size that would be functional and acceptable to a human astronaut.

Not all of the early space pioneers were human. When the Russians launched Sputnik II with Laika the dog aboard, a new dimension was added to space exploration. The world's physicists and engineers had been awed by Sputnik I, the first successful launching of a satellite into orbit. However, the biologists, flight surgeons and other life scientists were even more stunned by Sputnik II and its living, breathing passenger from the planet Earth.

And a new term was added to the national vocabulary: biosatellite.

There were many unanswered questions about how and whether a living creature, which requires breathing air and other life-supporting amenities, would fare in the strange, mysterious, unexplored environment of outer space.

There were two schools of thought about animal participation in the space program. The test pilots and aerospace scientists at the newly-formed NASA wanted to go directly to preparing human astronauts. But the flight surgeons and other medical researchers argued that animal studies were needed to safeguard the human pioneers.

The questions centered around life support. The planet Earth has its own life-support system which is largely taken for granted. No one knew if a living creature—of any species—could survive the stresses of launch and reentry, the enormous and sudden temperature fluctuations, and the exotic environment of weightlessness.

Surviving the initial stress and exposure was one thing, adapting to the weightless environment sufficiently to work, was quite another.

Packing for a trip into space would require attention to detail to a degree which had never been seen before. And much of the required life support system and biomedical monitoring devices had not even been invented. It was not known whether a life-support system could be successfully duplicated, carried into space and sustain life in the flawless way required for survival. The life scientists maintained that the need for animal explorers was twofold: to test out the many questions of trying to sustain life in such a hostile environment, and to be prepared to train animal astronauts to actually serve as pilots if it were deemed that space travel was incompatible—or inadvisable—for human astronauts.

This plan was vigorously rejected by the Mercury astronauts and their fellow test pilots. Indeed, they were scornful and insulted at the whole idea of animal astronauts. They referred to Sputnik II as "Muttnik."

One of Randy's friends, an Air Force colonel, had the misfortune of suggesting that chimps could be trained to pilot fighter aircraft for dangerous missions. His superior officers were so insulted by the mere suggestion that they transferred and removed him from the Air Force and retired him. To this day, the animal astronauts—and animal research—are not often mentioned in astronauts' autobiographies. The project went forward despite the critics.

Once the idea of animal research was accepted, the next question was, "which animal?"

Russia, with its tightly-controlled society, could launch a dog into space without fear of public reaction. However, in outspoken America, where animal rights groups are very vociferous, the scientists understandably shied away from antagonizing these groups.

The opposition to space research already abounded, so they reasoned there was no point in adding fuel to the fire.

Dogs—and also cats—were ruled out as biosatellite passengers early on, since millions of Americans are very sentimental about their pets. Rats and mice—at the other end of the popularity scale, were selected, equipped and trained. They were to be strapped into tiny space capsules and launched in biosatellites.

Some biologists recommended the tiny pocket mouse as an ideal space traveler. It survives in the desert with very little water. In a pocket mouse experiment, the tiny animals were outfitted with protective "clothing" and placed in miniature space capsules. However, before they were launched, the project was cancelled for reasons known only to the bureaucracy, denying them a place in space history.

It was finally decided that monkeys—and their cousins, chimpanzees, would be good subjects. They are shaped more like humans than many other mammals and they are also high on the intelligence scale.

The animal research program had many obstacles. As soon as one was removed, another surfaced. At that time there were only a few biomedical research programs involving monkeys in America and they were being used in natural habitat behavioral studies. Monkeys are not native to North America and most people had never seen them anywhere except at the zoo, where they entertained grandly with their antics. A group of monkeys was imported from India for space research purposes. They were placed in a colony at Holloman Air Force Base in New Mexico where they spent six months becoming acclimated. They were very well cared for. They had their own special veterinarian and as many comforts of home as could be managed. While this was taking place, scientists at the Naval Air Development Center in Pennsylvania, under Randy's supervision, prepared to conduct acceleration studies of primates on the centrifuge. At about this time, the Indian government began to complain about using monkeys—which have a religious meaning to them—for experimental space research.

This obstacle was resolved when the secretary of state smoothed the situation over and the American government decided to start its own primate colony so that no more monkeys were imported.

And so eventually, the chimps—accompanied by their personal doctor—began arriving at the acceleration laboratory to participate in the centrifuge flight training program.

The chimps received a lot of attention, even though at that time few people, including me, understood the significance of what they were doing.

One night we received a telephone call from the Memphis airport. It was the chimps' veterinarian and he had a message for Randy, "I'm here at the Memphis airport with the chimps," the veterinarian told me. "The airport is fogged in, so we're going to spend the night here and fly to Philadelphia in the morning." I could not help myself! In my mind's eye, I kept conjuring up a picture of the chimps, with small overnight bags, sitting patiently in the waiting room of the Memphis airport! The acceleration laboratory really didn't need any more commotion. However, the arrival of the chimp "astronauts" enlivened the place even more. With their entourage of caretakers and equipment, and their natural antics, they gave the lab the atmosphere of an exotic circus.

What was really funny—to everybody except the serious researchers—was the merry chase which ensued when they escaped their captors and ran wild—with their caretakers in hot pursuit. Sometimes they simply

The Chimpanzee selected as the spare or back-up animal for possible flight aboard a production version of the Project Mercury spacecraft on a Redstone-boosted ballistic trajectory is a female, approximately 36 inches tall, weighs 47¾ pounds, is approximately 4 years old and was born in French Cameroon, Africa. The animal was selected from among a group of six specially trained chimpanzees during the night before launch. She will be substituted for the primary flight animal in the event the primary animal cannot perform the mission.

escaped their cages; other times they were "helped" to freedom by some young assistant, eager for a break in the routine.

The chimps were accompanied by specially-trained animal handlers. They were airmen and Navy corpsmen whose job it was to help the chimps adjust to their space research role. Chimps are three times as strong as humans, and they could demolish a restraint system or other equipment if they were frightened.

To prevent this sort of mayhem, a bizarre "on-the-job" training program was developed for the chimps. The humans often climbed into the cage to reassure the monkeys. They also helped introduce them to their special food and also their new equipment in much the same way as new parents work with their infants.

The chimp research project was serious—even though it was so one-sided. The scientists planned each detail of the acceleration study with great care; the chimps remained carefree.

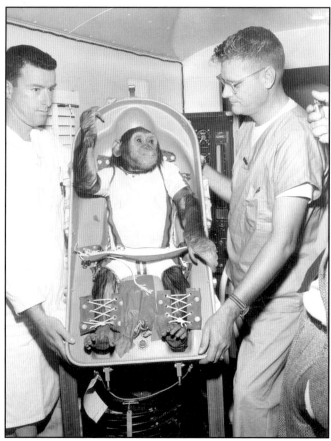

Chimp astronauts played an important role in helping scientists determine if living creatures could actually survive in outer space. This chimp astronaut is shown in a miniature contour couch which was launched into space.

The highlight of the chimp project was an experiment designed by Randy in which he could compare his own ability to operate the controls of a spacecraft with those of a chimpanzee.

Randy and Ham, the chimp, were seated side by side at twin control panels. They settled down for the 5½ hour test, surrounded by a crowd of Randy's co-workers and laboratory visitors. They started out as equals—more or less, but by the end of the test it was hard to tell who was studying whose problems. The chimp, appearing fresh and relaxed, not only held his own during much of the experiment; he even excelled in parts of it. Unquestionably, the chimp had some advantages. He came to the test after a restful night and a perfectly-balanced diet. Randy, on the other hand, had to worry about all of the problems and details of the experiment.

And it may be that the chimp's habit of letting out a bloodcurdling scream when he lagged in the contest, and of clapping his hands enthusiastically when he was ahead, was a bit disconcerting to his human competitor.

After the experiment, I asked Randy what he planned to do with the data he had obtained.

"I'm going to publish it in one of the scientific journals after we do more testing," he replied.

"Even the part in which the chimp excelled?" I asked incredulously.

"Certainly" he said, "That's the most interesting part!"

I had no doubt that this story would get back to our neighbors. However, Randy had never concerned himself with what anybody thought of his projects—and the boys and I had finally resigned ourselves to the fact that we were a part of that odd, but interesting, group of individualists dedicated to exploring outer space—whatever it takes.

A chimpanzee especially trained for flight in Mercury-Redstone 2 is shown in the suit to be worn by the monkey on the 115-mile-high, 290-mile-downrange ballistic flight. The chimps were trained by Aeromedical Field Laboratory, Holloman Air Force Base, New Mexico. MR-2 is one of a series of launches in preparation for manned orbital flight in NASA's Project Mercury program.

(Below) Evaluating a Man-Chimp display and response panel for an operant conditioning experiment in the AMAL Human Centrifuge is Dr. Randall Chambers, Head of Human Factors. Using identical panels and centrifuge exposure profiles, and similar couches, harness, and sensors, Dr. Chambers will compare chimp and human conditioning, G-tolerances, physiology, performance, and monitoring.

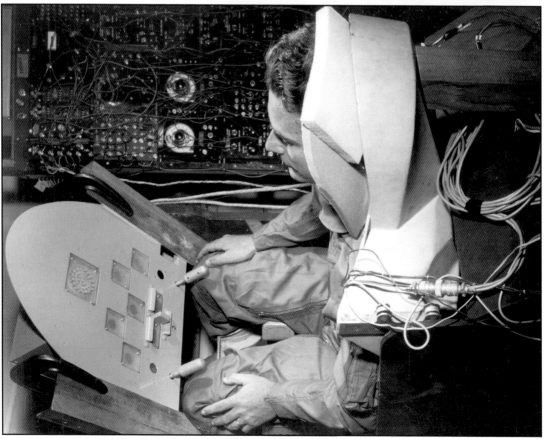

As it turned out, the chimps were intelligent and capable, but not committed to space exploration. One might say that the chimps were astronauts with an attitude. Whereas human astronauts worked very hard at training and were very pleased at being selected over great competition, the chimps were unimpressed by the honor of being blasted into space on a rocket.

Before one flight, Ham was given a banana and then launched for a practice run on the centrifuge. The next time he was offered a banana, he refused it. For a while, the group of experimental chimps led the life of the jet set, whether or not they were aware of it. One such instance was a trip Randy made, accompanying the group from Philadelphia to Washington. They were treated as the special animals they had become.

Randy, the monkeys, and the crew were all settled in a military airplane, ready to take off, when the crew received some chilling news: an admiral had decided to make the trip with them. He didn't know about the monkeys: he was just in a hurry to get to Washington on the first available flight. The crew became concerned that the odor of the monkeys might offend the admiral. So they rushed out and bought some floral deodorant and sprayed the unsuspecting creatures within an inch of their lives. By the time the admiral boarded the plane, the monkeys were as fragrant as flowers in a garden. The trouble was that the deodorant almost asphyxiated everyone aboard, including the admiral.

The final blow came at the Washington airport where a crowd of press photographers awaited the plane. The admiral alighted, poised himself and his gold braid ready for the news photographers and the television interviewers—only to see them rush by him on their way to photograph and interview the monkeys.

Experimental chimps were more of a novelty than admirals!

Primates were used to study various aspects of space flight for several years. Among the most significant projects were studies at Pensacola, Florida and Yerkes Laboratory in Atlanta, to study the effects of prolonged confinement on the body. Chimps were confined to bed for a month and as a result suffered skin problems, muscle weakness, bone loss, blood chemistry imbalances and other effects. These early studies resulted in a complete about-face for surgical patients and others. No longer are patients allowed to rest in bed as had been the case for years. As many patients have learned, almost immediately after surgery, the patient is required to get up and move around. Even geriatric patients are assigned to a physical therapist who arrives at the patient's bedside with a set of prescribed exercises.

As a result of these studies—and others—astronauts on long space trips also were equipped with exercise facilities and fitness routines.

Our sons were growing up with the space program and it was influencing their thought processes. Mark, the oldest, began browsing through his father's books as a pre-schooler. One of his favorites was a little zoology text which had pictures of animals according to their biological classification. One day he opened the book to the primates and asked me to explain it.

So I told him that man was at the top of his family tree with apes, monkeys and baboons on the lower limbs.

"What's man doing up there above the ape?" the child asked. Sometimes I think this is still a good question! Randy—and many of the other space researchers—were very dedicated to the task and give the impression at least that no sacrifice is too great for the cause. I'm sure that Randy—as well as some of the others—did dangerous things, such as riding the centrifuge at higher levels than were deemed safe.

I didn't find out about a lot of these daring feats—they kept them to themselves like a group of school friends gathered in a tree house, and swearing each other to secrecy.

In the midst of the animal studies I did, however, discover that there was a plan for Randy to accompany a black bear on a flight to New Mexico. I reacted with my usual borderline hysteria. Was the bear fresh out of the wilds? Was he going to be caged during the trip? Was he at least going to be tied up? Was he going to be well-fed before boarding the plane? Did he like people? As friends or as a main course?

Randy hadn't bothered to find out the answers to any of these questions, which I considered to be vital. I thought flying across the country by jet was hazardous enough in those days without giving your hand to a bear who might not give it back. In the midst of my nagging, the trip was cancelled. Not because anybody in the government thought it was dangerous, or because a local group demanded that a western bear get the job. No, the whole thing was called off because the bear wasn't up to making the trip!

However, bears were still being considered as possible space travelers in addition to the chimps. Bears were a favorite of some scientists because their size and shape were more similar to humans.

Another early candidate for space travel was the pig. Physiologists pointed out that pig's skin and flesh is very much like that of humans. For early studies, a contour couch and other equipment were designed for a pig. The pig was disqualified when it was discovered that it could not breathe freely while lying on its back in a launch position.

Eventually, however, the only non-human to finally be launched experimentally into space by the U.S. were the monkeys and chimps. Ham, Randy's competitor in learning to operate the controls of a spacecraft, was the first to be launched. The earliest attempts at launching animals in rockets in this country took place in 1928. The experimenter was Dr. Robert H. Goddard, an American physicist. (NASA's Goddard Space Flight Center near Baltimore, Maryland, is named for him.)

Dr. Goddard attempted to launch chickens in rockets to study their survivability. In those days rockets were in the early stages of development and were similar to cannons and firecrackers. A missile does not become a satellite until it is launched into orbit. No one succeeded in launching a satellite until the Russians put Sputnik I into orbit in 1957.

Animal research—conducted over the years—also has made some unexpected contributions to the space program. For example, one of our friends in Maine was studying aggressiveness in goats in the 1950s. He was trying to determine which goat was the leader of the herd by recording their butting behavior. This went on for weeks. In all kinds of weather, he was out there in the field, notebook in hand, watching the goats butt each other and lock horns. The scientist looked like a referee in some sort of peculiar boxing match.

At the time, I wondered what use this kind of information could possibly have. Once you find out that Brownie is the undisputed leader of the herd, what can you do except congratulate him?

Years later, however, space scientists were combing the scientific literature for every scrap of data they could find about aggressive behavior. After all, personality characteristics such as aggressiveness become very important to astronauts who are confined together in small spaces for long periods. Another friend of ours studied the effects of isolation and confinement—another aspect of space flight. This scientist used a dog as a subject.

The problem was that dogs are friendly animals who refuse to be isolated. The dog was presumably being raised without any contact with humans. It had been carefully isolated since birth and its physical needs were taken care of by mechanical means. But every time the researcher peered into the enclosure to observe the dog, his gaze would be met by a pair of velvety brown eyes peering back at him and looking as though he were waiting for a chance to lick his face.

Another dog was to be tested by a maze—a long, complicated winding passageway lined with mirrors. At the end of the maze was food, and the purpose of the whole thing was to see how quickly the dog could learn to find his way through the maze and get to the food. The scientist brought the hungry dog in, deposited him at the starting point of the maze, and shut the door. He waited a few minutes and nothing seemed to be happening. The dog was supposed to be working his way through the maze, but all seemed to be too quiet. Finally, the scientist opened the door and there was the dog, sitting at the starting point—waiting for somebody to bring him his dinner. The expression on his face said that he had no intention of going through that maze just for an ordinary dinner of dry kibble.

During the early years of astronaut training, our family acquired a pet cat. Actually, it was Craig's idea. The cat who lived next door had produced a litter of kittens, which was really none of our affair. But Craig paid daily visits to see how the brood was getting along. Both Craig and I knew that Randy much preferred dogs to cats, so I didn't think Craig's campaign to adopt a kitten would have much success. But Craig was an experienced psychologist, even at the age of four. He didn't come right out and ask for a kitten. He gave us a choice: he said he wanted either a kitten or a baby brother. We adopted the kitten the next day and Craig named him "Dennis."

Even as a kitten, Dennis immediately accepted the boys as full partners. When they played jungle safari, Dennis allowed himself to be "shot" by a toy gun, thrown over a small shoulder, and lugged back to camp as the tiger of the day. The boys could pick him up out of a deep sleep and he was still happy to be their lion, tiger, or whatever role they chose for him. He even let them drag him into the coat closet and study the pupils of his eyes with a flashlight. And he acted as though he enjoyed it when they stroked him hard to see how much static electricity they could generate. He seemed to feel that these things were a part of his job.

But even as a kitten, Dennis let it be known that he had no intention of becoming a space cat. We acquired him during the early Mercury chimp flights and our house was a gathering place for space scientists, test pilots, astronauts, aeronautical engineers, flight surgeons and medical monitors. Randy had only to look in Dennis' direction and make some offhand remark to get a glare from the cat.

"Look at those wonderful reflexes," he would say. "My what good reaction time that animal has!"

Or, "Look at that cat—he's a marvel of domestication. Did you ever see anyone who could adapt himself to his environment so well?"

Dennis reacted to this kind of talk with a look which said, "You use your brain and I'll use mine!"

The cat also gave a wide berth to any of our visitors who showed the slightest interest in his reflexes. He often simply hid under Craig's bed the minute guests arrived.

Ham's historic ride into space took place in 1961 when he was launched from Kennedy space flight center in Florida into a 5,000 mile-an-hour ride down the Atlantic missile range.

By this time it had been established that humans would survive space travel with reliable vehicles, well-designed, protective clothing, improved communications, and much careful training. The emphasis was then placed on human space travelers and the chimp program wound down. However, chimp and other animal studies continued to enable scientists to understand the effects of long-term isolation and confinement in the microgravity environment.

The big question was whether to build devices to provide artificial gravity for long term space travelers. It was not known whether the human body could survive and function without its customary (and essential) gravity, food, water, and breathable air for a prolonged period.

It was one thing to make a brief trip into outer space but quite another to stay for an extended period. Monkeys were then used to study these long-term effects. These studies continued until the 1970s in preparation for the launching of Skylab, the U.S. space station. Ham and his relatives had made their contribution.

When the first astronauts dazzled the world with their unbelievable flights, they became the heroes of the hour. Theirs looked like a glamorous life—a sort of combination of super-pilot and explorer of the heavens combined with rock star fame. Many of the early flights were concluded with visits to the oval office of the President and ticker tape parades. However, few people—even the space aficionados— have stopped to wonder how the role of the astronaut came to be defined and how the astronautical job description and requirements were developed.

The basic question for developing criteria for astronaut selection and training was like one of those riddles about which came first, the chicken or the egg. In the 1950s there was little astronaut training data in existence.

The question was how to select and train astronauts to travel into an unknown environment which had never before been visited by humans. And to determine what kind of equipment they would need to do something that had never been done before.

It soon became clear that the process would evolve step by step as more was discovered about the physiological and psychological limits of the human body as well as the mysterious space environment.

Added to this puzzle would be the evolution of the equipment, most of which had to be developed from specially-compounded materials after thorough research.

The entire effort was conducted by the uniquely American system of bureaucracy, politics and capitalism.

The first astronauts were all experienced military pilots. They were recommended by advisory groups such as the National Advisory Committee, the Office of Biotechnology and Human Research, the Office of Manned Space Flight, the NASA Astronaut Selection Board, the National Academy of Sciences, the Office of Space Science, or one of various other human factors, life science, biomedical, behavioral sciences, psychological disciplines and engineering selection boards. All were a part of the sprawling federal government system and its contractors. In other words, the military pilot pool was recommended by a committee—probably one of the biggest committees anywhere. Also, a number of space committees from the Department of Defense offered recommendations for selection and training of this elite group.

One of the major criteria was to be a highly-experienced test pilot who was also knowledgeable in aeronautical and space sciences. Also included was willingness, motivation and determination to undergo the stresses, strains and hazards of space flight and training. Many of the hazards were unknown or even ill-defined.

A list of major hazards is long: high gravitational forces with spins, turbulence and impact in fluctuating mission profiles, during launch and reentry; physiological effects of prolonged isolation and confinement; possible fire and explosions; depressurization of the spacecraft; hazards of escape and flight abort conditions; possible radiation exposure and injury; effects of wild temperature fluctuation; oxygen deprivation and breathing air emergencies; physiological and psychological problems of functioning in microgravity; excessive pain and discomfort inherent in emergency situations and concern about defects or failure of systems and equipment while in flight. Their most prevailing fear was of a catastrophic crash due to pilot error.

These were not unfounded fears. At that time there had been numerous equipment failures—including a high percentage of launch system failures. There was also the question and concern about how the spacecraft—and the pilot—would perform in the real space environment. No system had ever been tested in space and no pilot had ever ventured into space.

Designs on the drawing board and tests in the laboratory don't always succeed when put to the real-life test. The fact that space is such a dangerous, unfamiliar environment added to the anxiety of both the astronaut candidates and the scientists. Despite these drawbacks—which would no doubt be a deterrent to sensible people who enjoy life in 1G gravity, there was great competition among the military pilots to join the program. There were more than 900 volunteers vying to become the first U.S. astronauts.

There were no standardized tests nor job descriptions for aspiring astronauts. Also, there was no central recruiting office. The 900-plus candidates were screened and gleaned in widespread places by a variety of officials, panels, and committees. The military hierarchies also played an important role in astronaut selection. There soon developed a power struggle between the infant NASA and the Federal Aviation Administration, some officials of which felt that the entire project should have come under their supervision in the first place. Reasons for being screened out of the competition were numerous and varied. Many were eliminated early for such reasons as not having enough flying experience. However, in what appeared to some to be a reversal of this criterion, was the fact that age was also a factor. There seemed to be a delicate balance between flying experience which was highly valued and becoming too old in the process of acquiring this experience.

Size and body shape were also an important considerations. Being tall is a requirement for becoming a basketball player, but being short was much preferred for would-be astronauts. Being small boned was also an advantage. Successful candidates would find themselves working in very small space compartments. Education—especially degrees in engineering and science—were also a plus, as well as having graduated from one of the prestigious universities.

There were other more subtle reasons for eliminating prospective candidates. The military services vied to have a balance of different branches of the services represented. Admirals and generals sponsored their favorites. Some top-ranking leaders took the acceptance—or rejection—of their candidate personally. Luck—being in the right place at the right time—was also a factor. Finally the first astronauts were selected: the seven Mercury astronauts. Three were from the Navy; one from the Marines and three from the Air Force. There was a great furore about the fact that no pilot from the Army had been selected.

The original seven NASA Astronauts: Alan B. Shepard, Jr., Walter M. Schirra, Jr., John H. Glenn, Jr., Virgil I. Grissom, M. Scott Carpenter, Donald K. Slayton, and Leroy G. Cooper, Jr. They autographed this picture, "To Randy Chambers with sincere appreciation for your help."

There were also many civilians from various walks of life clamoring to be astronauts. This was especially true of scientists, engineers and commercial test pilots. There were also applications from pilots from other countries.

One civilian test pilot with a science-engineering background and experience in testing the X-15 rocket ship and the Mercury centrifuge projects was a young man named Neil Armstrong from NASA Ames and Edward Air Force Base. Randy had worked with him on the X-15 and other research projects. Neil had also helped develop some of the training programs for Mercury, Gemini and Apollo and Randy had been impressed with him. (As the world knows Neil was selected later for the Gemini series and of course became famous as the first man to walk on the moon.)

There were a number of women who also felt left out of the astronaut selection process. Some even went so far as to demand a hearing from the U.S. Congress. Like some husbands, Congress listened politely and then tabled the issue—permanently. A few women even succeeded in getting a chance to ride the human centrifuge at the naval aviation medical laboratory. However, the medical directors of the laboratory refused to allow women to go beyond 6-G tests, despite their protests. (This was an era when school girls played basketball on only half a court because females were considered to be fragile creatures.) The whole system of astronaut selection has since been refined to include international participation and a broader selection of both men and women. The crews were also broadened to include not only pilots and astronauts but mission specialists and other specialists.

The selection of the astronauts was just the beginning of the process. When, where and how to begin to design and set up a training program was an even more complicated undertaking.

In later years the term "the right stuff" was used rather glibly; the fact of the matter was that designing astronaut selection and training criteria was no small task. The major obstacles to getting off the planet were the physical forces of gravitational stress, atmospheric pressure which affected both man and machine; and microgravity (weightlessness). Other hazards, such as electromagnetic radiation, the effects of isolation and confinement and chemical contaminants, and other environmental challenges were also addressed.

Those of us who are non-scientists would probably have tried to fight the physical forces. However, scientists are serious students of physics and they know how unbending these forces are. The scientific approach was to set up training programs to prepare the astronaut to accommodate to gravitational stresses, and also to design and outfit him with protective clothing and equipment. The astronaut also had to perform piloting tasks under these strange new flight conditions.

The problem was solved in increments, using the age-old idea of starting with what is known and then gradually moving into the unknown by carefully testing, monitoring, evaluating, and checking of their responses throughout the simulated flight. Dr. Chambers designed many of these studies. He began with what he knew about the human body and brain and then began testing their capabilities on the human centrifuge. The giant simulator created gravitational stress and could be adjusted to many different profiles. These basic studies were tried out by scientists, engineers, test pilots and other volunteers who rode the centrifuge with great dedication and boundless enthusiasm for the exciting project of sending astronauts to the moon. From the data they collected, the centrifuge training programs were designed and ready for the first astronaut candidates.

Atmospheric pressure, which is vital to the functioning of both body and machine, was studied in pressure chambers. Sometimes the centrifuge gondola itself was used as a pressure chamber. By the time the astronauts were trained, equipment had been designed to allow the space men to travel into space in an envelope of vital atmospheric pressure. Weightlessness was also simulated, mostly in water tanks which provided neutral buoyancy. The water tanks were the site of many hours of practice for understanding and functioning in weightlessness.

Dr. Chambers and others designed astronaut training programs for these simulators as well, using the known-into-the-unknown methods of testing, evaluating, monitoring and checking.

In addition to developing an astronaut selection and training program, the vast array of materials and equipment needed to carry the astronaut into space and return safely also had to be developed and tested.

In the late 1950s the aviation medical acceleration laboratory with Dr. Chambers heading up the acceleration training programs, took on the appearance of a center for discovery and exploration—a kind of half-way station between Earth and outer space. Things which had never been seen outside of Buck Rogers comics or science fiction movies began appearing at the laboratory contour couches, pressure suits, intricate helmets and high-tech equipment that required huge bundles of wire.

Another major part of this new space adventure was the rocket which would carry the vehicle into space. At the time the astronaut selection began, rocketry had to be redesigned to carry human cargo. In fact, many scientists—and non-scientists as well—felt that the rocket was the most untrustworthy of the entire nerve-wracking arrangement. The design of a space vehicle which could be placed on top of a Redstone or an Atlas rocket had not yet been completed. A training program to prepare the astronaut to control the rocket during the crucial launch also had to be added to the vast preparations. There were many critics of this plan who felt that placing a human being on top of a powerful rocket was simply too dangerous. At that time there were many proponents of other types of vehicles. There were three basic vehicle designs—the glide vehicle, the drag vehicle which was being developed for the Mercury, Gemini, and Apollo flights, and a combination vehicle. They were named for the way they performed. The glide vehicle, which later was used for the shuttle program, had wings and glided into space. The drag vehicle, used for the earliest flights, was designed like a funnel-shaped container and dragged along much less gracefully. The training requirements were quite different for each vehicle.

One of the first discoveries of the centrifuge high-G experiments was that at certain G-levels the pilot would be unable to talk, to see, to operate control devices, to breathe or to maintain consciousness. His heart, circulation, neurological and biophysical functions were also of concern.

The original military pilots selected to be NASA Astronauts. Left to right: Lieutenant M. Scott Carpenter, USN; Captain L. Gordon Cooper, Jr., USAF; Lieutenant Colonel John H. Glenn, Jr., USMC; Captain Virgil I. "Gus" Grissom, USAF; Lieutenant Commander Walter M. Schirra, Jr., USN; Lieutenant Commander Alan B. Shepard, Jr. USN; and Captain Donald K. "Deke" Slayton, USAF.

These factors were all-important for a safe space flight. So it was apparent that the spacecraft had to be designed to be operated at levels below these performance thresholds. It was also discovered that practice at high-G on the centrifuge could bring about training and conditioning so the pilot could perform at higher levels for longer periods of time. It also helped familiarize the pilots with the effects of acceleration forces. Centrifuge training was so successful that astronauts returned for retraining before each flight.

Even though all of the successful candidates were military pilots, each branch of the service trained their pilots differently. However, all had been trained on flight simulators and they were all highly trained in military discipline.

All of the candidates had been flying airplanes at low-G levels. Now, one of the first requirements to become an astronaut was to tolerate the effects of high-G conditions on pilot performance far beyond their previous experience.

Now they faced many factors that were unfamiliar to them. These included preparing for the stresses of high-G launch, reentry and impact; learning to work and survive in the exotic environment of weightlessness, and surviving the hazards and sensory deprivation of prolonged isolation and confinement.

A spacecraft is very different from an airplane and the astronaut candidates set about learning to operate the controls of this unfamiliar new vehicle and to do so in the unusual space environment. This was especially true during training for emergency and abort conditions. Communications systems were far different from any they had experienced earlier. There are black-out periods of silence during space communications and the hardest lessons were to learn to adapt to them. They also had to practice the art of adapting to the space environment in order to focus on accomplishing mission tasks such as taking pictures and collecting data about outer space.

Another obstacle which loomed large was learning to tolerate the unusual changes in atmospheric pressure which they would encounter for sustained periods of time. They had to train their bodies to tolerate 100% oxygen in their breathing air system. Working with 100% oxygen also created fire hazards which in turn created still more training to understand the safeguards.

They all had undergone survival training in their military careers, but survival training for space was a new challenge. They had to practice for unexpected catastrophes such as fire, crash landing, or being stranded in unexpected places. Special pressure suits and other items of protective clothing were far different from anything they had experienced previously. Aviators dressed themselves. The first astronauts had to be dressed by others. (Much later, simpler suits were developed.)

The new astronauts also had to practice walking at $1/6th$ gravity to prepare for moon walking. This was a bizarre exercise practiced on a lunar lander simulator which itself looked like some oddity from another world.

Early biomedical monitoring was very cumbersome. Electrodes and wires were taped or strapped to the chest to monitor heart, blood pressure, respiration, temperature, brain activity and vision. This instrumentation package sometimes slipped off during training. Since continual medical monitoring was required, any test interrupted by the loss of the biomedical monitoring had to be restarted from the beginning. Patience became a basic requirement for this work.

There was also a moment of agony when the tape was removed from a hairy chest. Sometimes hair was shaved off for the placement of electrodes. (This ordeal also tested the astronaut's commitment to space exploration!)

The astronaut's ability to be patient and cool under pressure was tested by his reaction to the stressful conditions and the inevitable interruptions and delays. They were also being tested—sometimes surreptitiously—on a performance and evaluation ratings scale. This included the willingness and ability to perform the mission task on the exact schedule despite fatigue, discomfort, and annoying distraction.

The goal was to perform the task at hand—whatever it takes. Astronaut training lasted for several years and included familiarization, training and retraining in a variety of simulators and training devices in various parts of the country. The human factors part of the project—that is, the human-

Astronaut training and centrifuge simulations for Mercury, Gemini and Apollo missions always included extensive safety testing and evaluation before, during, and following the centrifuge flight simulations. In these early simulations the astronauts frequently tested components of their life support system while making comments about safety, performance and comfort.

machine aspects—was done by scientists and technicians in laboratories of the government, various industries, and universities.

Dr. Chambers' role was to coordinate this effort to produce successful manned flights. He participated in the design and research for many of the projects and reviewed many others.

The "right stuff" was a popular term for describing the qualifications for astronaut job descriptions, but determining what requirements should be included in the right stuff was a very meticulous, methodical process.

A simplified list of the steps taken for each increment of the research reads like this:

1. Investigate the problems for each task in the sequence.
2. Design tests and experiments of man-machine systems.
3. Conduct trial runs and test the design.
4. Analyze the data.
5. Design training programs and facilities.
6. Conduct the training.
7. Follow up with refresher training.

All of these steps were done in sequence and then each sequence was repeated for each new stage. For example, the Mercury capsule was designed for one pilot. The whole research sequence was then repeated for Gemini, which had two pilots and then again for Apollo, which had three. These steps were also repeated for sub-orbital flight, then orbital flight; weightlessness, orbiting the moon; landing on the moon, etc. These steps also were performed for each aspect of each discipline such as medical, psychological, engineering and flight dynamics. There was input from the astronauts who were asked for their comments and observations throughout the testing. They were interviewed intensively before, during and after each experiment. There were so many questions to be answered that one would hardly know where to begin.

The astronaut candidates who might be called "semifinalists" were tested and rated extensively before final astronaut selection or mission training could even begin.

The combination of new and unusual challenges was almost endless. The training program would have to include total simulation including pressure suits, breathing air equipment, visual acuity and display interpretation, depressurization, realistic pilot restraint system, display and event panels, control devices, pilot performance, physiological and psychological instrumentation, monitoring and communication devices, and rocket system controls.

The space men all wore contour couches which helped protect them from gravitational stress. Many trial runs were performed to study G-forces of various types. The training program was set up before the final selection of a rocket was made. Eventually, the Redstone rocket was used for sub-orbital flights and the Atlas for the orbital flights. Astronaut training for controlling the rocket varied greatly for each type.

Anyone looking at the controls for spacecraft simulation saw more equipment than anyone could possibly imagine—or use. There was even an "event" panel which displayed the activities of all the other panels.

This describes an enormous effort which was, in all probability, the largest cooperative interdisciplinary science project of the 20th century and perhaps of all time. But no one—with the possible exception of science-fiction writers—has ever said that getting off the planet would be easy! The space program increased the interest and effort in the study of science and engineering at all levels of education. School teachers, college professors, and broadcasters of all kinds were forced to learn more about such things as aerodynamics, physics, astronomy, mathematics, physiology and other scientific subjects so as not to appear ill-informed.

The second group of NASA astronauts were selected for training and flight in the Gemini and Apollo missions. The original seven Mercury Astronauts were included in this selection of sixteen for Gemini and Apollo.

Astronaut Donald K. Slayton, riding the human centrifuge to simulate Mercury acceleration G-forces for familiarization and training, observes and operates some displays and controls in the cockpit. The astronauts were given extensive G-experience and practice in controlling their simulated spacecraft during launch and reentry simulations in this cockpit used as a training device.

For his part, Dr. Chambers became one of the world's authorities in human performance in unusual environments and also in space flight simulation. He received numerous awards, honors, and medals for this work.

Astronaut training was the reversal of most programs in that passing the first round only led to more rigorous requirements for the next round. Those who showed the ability to endure pain and discomfort were selected for more and more intense pain and discomfort. My memory of them is that they were young, healthy, vigorous, enthusiastic and as brave as anyone could have expected. Even though the astronaut candidates who were chosen were so thrilled to be included in such an elite group, they also had their complaints and their misgivings. Many were heard to grumble—only half-jokingly—that they were entrusting their lives to equipment produced by the lowest industrial bidder. There was also grumbling about those "NASA doctors and experimentalists" about some of the tortuous research projects they designed.

"Was this really necessary?" was a question heard often. Although most of the would-be candidates who were eliminated from the competition were screened out by one official or other, some hopefuls dropped out voluntarily. They found the testing, probing, poking, prodding, analyzing and questioning intolerable. The most amazing thing about this early stage of astronaut selection and training was the number, enthusiasm and determination of candidates willing to sit on top of a powerful rocket with a 1961 failure rate of over 60%. The tumultous system of developing the requirements for "the right stuff" did not dim the excitement and adventuresome spirit for space exploration.

Space flight was a test of human perseverance, endurance and survivability. But even more challenging were the physical laws which heretofore had been encountered mainly in physics textbooks. Gravity, the atmospheric pressure factor and weightlessness had all been challenged at various times by acrobats and daredevils, but rarely to the extent required to get human beings off the planet.

One of the biggest obstacles of all for the human pilot was gravitational stress which presented serious problems for space flight during launch; reentry and planetary landings.

Astronauts had to be trained to perform piloting tasks and manage their life support systems while accommodating to the physical forces. Any gravitational loads other than the 1-G enjoyed on Earth created problems for space researchers as well as space explorers. One must exceed the pull of gravity to get off the planet—even by powerful rockets. It is impossible to get off the planet without exceeding the force of Earth's gravity. Achieving a speed of at least 17,000 MPH at an altitude of more than 100 miles is necessary in order to continue the escape from Earth's gravitational pull and to stay in orbit around the Earth.

Earth's gravity is so strong that it could pull a vehicle out of orbit and back to Earth if the speed isn't maintained or exceeded. After the vehicle orbits the Earth it has to pick up speed in order to travel to other planets such as the moon. As the vehicle approaches another planet, it will be influenced by that planet's gravity. While preparing to land on the moon, the astronauts practiced walking at $1/6$ gravity in a lunar lander simulator in order to be prepared for lunar gravity. Pilot performance has to be adapted to the changing gravitational requirements.

On the return trip, the astronaut and the vehicle have to adjust to the gravitational changes in reverse. Gravitational stress is a challenge in both leaving the planet and in returning to Earth. Penetration of Earth's atmosphere is a danger zone full of such perils as extreme heat, extreme cold, and extremely high-G forces.

This danger zone has created many hazards for both astronaut and vehicle which have to adjust and withstand such wildly fluctuating conditions. The G-forces encountered in launch and reentry are produced by the resistance of the Earth's atmosphere. Earth tries to keep everything held tightly in place. It resists mightily anything that tries to come and go. Most of us Earth dwellers appreciate the fact that we don't have to worry about accidentally falling off the planet. But the problems are many and varied for those who are trying to leave or return.

Acceleration forces vary in their amplitude, direction, duration and profile, occurring in variable sequences. This might compare to being caught in a tornado. This means that to prepare pilots and vehicles for space travel the scientists had to study the effects of various acceleration forces such as linear, centrifugal, radial, angular, impact, turns, rotational, vibration, orbital, spins, oscillations, egress, tumbling or ejection.

The astronaut had to be protected from these forces and trained to guide his vehicle through these changing, sometimes unpredictable, G-forces. Included in the studies were gravitational forces occurring during escapes from spacecraft, splash down, emergency, or crash. And after conquering every acceleration force imaginable, they had to reverse the process and study deceleration forces in order to slow down and reenter the Earth's atmosphere safely.

The best example of assorted gravitational forces on Earth can be found in amusement park rides. Ferris wheels, roller coasters, merry-go-rounds and assorted airborne spinning and swinging cars give fans a sampling of various gravitational forces and their effects. Very early in the space program, amusement park rides were considered as possible research vehicles to study acceleration forces. However, it was soon decided that those rides were not inherently strong enough to be used for the extreme stresses that were needed for a comprehensive study. (Fans of newer, more daring roller coasters would probably disagree!)

This modified version of the human centrifuge was equipped with a new gondola, larger than earlier models. The gondola could be pressurized and could be rotated at various speeds in pitch, roll and yaw angles. This new gondola enabled scientists to study several angles and speeds of acceleration at once. This is clearly no place for anyone with a tendency for motion sickness or disorientation!

The study of acceleration stresses was such serious business because the effects of these stresses on the human body—as well as their equipment—are so life-threatening. Unless properly prepared and protected, high acceleration forces cause the eyes to stop functioning which causes gray-outs, black-outs and disorientation and then unconsciousness. The brain, with its cognitive functioning and multitudinous capabilities also had to be protected. The heart, the cardiovascular system, the respiratory systems and all of the other organs are also seriously affected by gravitational stresses. Equipment—if not made of the right stuff—can also be affected by gravitational stresses, which can cause it to melt, tear, disintegrate or malfunction. One prime example is the heat shield protecting the spacecraft. This has been—and still is—one of the most difficult and recurring problems. Space scientists of different disciplines varied widely in their view and their studies of G-forces. The engineers studied the effects of acceleration stresses on the vehicle and the other equipment. The life scientists studied the reactions of physiological and psychological effects of gravitational stresses on a human, particularly the major physiological systems.

The fact is that the vehicle and the other non-human equipment could withstand the stress much better than a human. It was clear from the beginning that acceleration forces could shake the human body apart like a rag doll in a hurricane if protective equipment was not designed and provided.

Complicating an already-complex situation was the fact that different kinds of G-forces affect the body in different ways. For example, the angular and radial acceleration forces produced disorientation and motion sickness. Linear acceleration may produce gray-out, black-out, unconsciousness or death due to failure of other organs.

Comparisons of contour couches used in early Mercury acceleration training and research at the Aviation Medical Acceleration Laboratory. These couches, left to right, were fitted for: Astronaut Alan Shepard; Dr. R. Chambers, AMAL Project Director, and Astronaut Virgil "Gus" Grissom. Physiological performance and acceleration effects were compared during testing.

There were always two major areas of concern—physiological tolerance and performance tolerance. The specter of an unconscious astronaut hurtling through space beyond the reach of ground control haunted the scientists for years. Physiological tolerance to G-stress in space was basic. Equally important was performance tolerance—the ability to perform piloting tasks while exposed to acceleration forces. Both were studied separately and together.

Criteria for physiological and performance tolerances were built into the spacecraft in order to protect the crew.

This meant that the rocket and the spacecraft system were designed after much research on human performance tolerance and physiological tolerance. This research was the key to the basic question of whether a human could survive and pilot a craft in such perilous and unusual circumstances. Even though they were all experienced, highly-trained military pilots, the new astronauts soon found themselves facing unfamiliar problems.

An astronaut would encounter much higher G-forces than an airplane pilot. The space vehicle, unlike the airplane, also had to be capable of withstanding much more severe flight conditions. This included sustained high-G forces at different stages of launch and reentry, sustained micro-gravity conditions, abrupt and extremely high G-forces of impact on splashdown and many angular and rotational combinations. There were people who thought airplanes could be modified to go into space. However, as research progressed this idea faded.

Using the AMAL Centrifuge as a space flight simulator and training device, Dr. Carl C. Clark prepares to evaluate some Mercury-type acceleration profiles and G-tolerance effects. He plans to operate some early concepts for cockpit panel displays and flight control equipment, and review pilot restraint equipment and locations of restraints, panels and controllers within the cockpit. In the 1958-59 time frame, seat locations with respect to the best G-tolerance for physiological and psychological functioning were being tested and evaluated for early Mercury concepts.

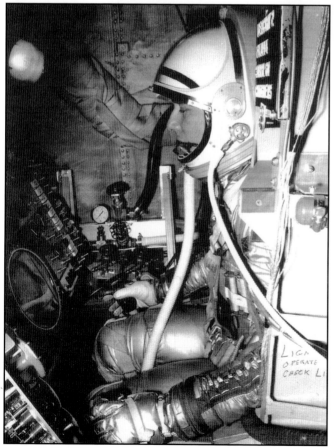

Dr. Randall Chambers, wearing a Mercury pressure suit, spent many hours in the gondola of the human centrifuge, studying the effects of acceleration stress in order to set up training and research programs for the astronauts.

NASA Astronaut, Alan B. Shepard, Jr., Lieutenant Commander USN, at the pilot loading platform of the Navy's Human Centrifuge at the Aviation Medical Acceleration Laboratory. Astronaut Shepard is preparing for a centrifuge simulation of the acceleration flight profile of a Mercury flight. Al spent many hours in research and refresher training in early centrifuge simulations of Mercury flights.

(Below) Astronaut Schirra, wearing his pressure suit and helmet, seated in his contour couch, observes and reports subjective acceleration effects as he flies through a centrifuge simulation of a Mercury flight. He flew missions on the Navy's human centrifuge, using the centrifuge as a simulator and training device.

Another factor new to the first astronauts was the rocket, an essential part of the spacecraft system, but a somewhat unfamiliar concept to a fighter pilot. At that time the rockets far exceeded the tolerances of human beings. The control system in particular had to be modified. Then a whole new series of rockets—the Atlas—was designed to replace the Redstone. Since the spacecraft had to separate from the rocket after launch, the concept that an airplane could be modified into a spacecraft died completely.

The space system required a large number of special performance tasks which airplane pilots had not been required to do. For example, the astronaut had to activate a booster cut-off switch which would separate the spacecraft from the rocket at a precise moment. And to prepare for landing, the astronaut had to operate the retrofiring mechanism which turned the spacecraft into position for beginning reentry. He also had to learn to manage the craft through the rigors of reentry.

Another difference which was unnerving at first was that, unlike an airplane, the spacecraft is cut off from communication for 4 or 5 minutes during reentry. After reentry the astronaut operated and monitored the drogue chute and main chute which stabilized the spacecraft as it splashed down into the ocean.

All of these tasks were new to even the experienced airplane pilots and had to be performed with great precision.

Rotational G-forces posed special problems because rotational stress brought with it illusions and false perceptions. Studies to overcome these stresses were conducted on a giant 75-foot centrifuge in Santa Monica, California. This enabled the scientists to study rotational stress during long, slow acceleration which would be encountered on extended space flights.

Performance tolerances for Gemini and Apollo centrifuge training were different from Mercury because of different launch, orbiting and reentry profiles. Transitions had to be made quickly. Not only did the body have to adapt to such stressful changes but it had to make quick transitions from high-G at launch to microgravity in weightlessness and then back to more acceleration stress for reentry.

One of the most memorable early centrifuge studies which Randy devised was "Eyeballs In, Eyeballs Out, Eyeballs Up, Eyeballs Down." When I first heard this I thought it was just a catchy title. I should have known better.

What this study entailed was having the subject ride the centrifuge at various positions which forced the eyeballs inward or outward during transverse acceleration; left or right during sideward acceleration loads, or upward or downward during positive or negative acceleration. Tolerances to these various positions in the head were carefully measured. The earliest astronauts might have needed to be able to perform piloting tasks at gravitational stresses reaching as high as 14-G for 5 seconds during launch. The extensive centrifuge runs for the early astronauts were designed to prepare them for accommodating to this acceleration stress.

Also included in the testing was a series of the sustained low-G centrifuge runs for 2-hour and 4-hour periods. Included in this series were test runs peaking at 3, 5, 7, 9, 11 and 13-G for varying intervals.

Using select positions in the human centrifuge for "Eyeballs In"-type profiles resulted in better pilot performance than for "Eyeballs Out or Down." The types of training usually provided on this centrifuge included environmental conditions, flight task requirements and procedures during exposure to the environment conditions and flight task requirements. In addition to the centrifuges, other combined gravity simulators were available. These included multi-axis rotation cabins, aircraft simulators, rotating rooms, air bearing frictionless devices, vibration platforms, aircraft used as simulators, neutral buoyancy simulators, rotation devices in atmospheric pressure chambers, and a variety of gravity, visual and noise simulators. Special protective equipment was developed for protecting the eyes and hands from "Eyeballs Down and "Eyeballs Up" maneuvers. In all cases over 3-G there were performance decrements as a function of time.

Astronauts had to be trained to perform piloting tasks during all phases of space flight so learning about gravitational stress and its multitudinous effects became an essential part of the training program. These centrifuge studies resulted in the development of 24 basic principles for determining tolerance levels for different amounts, time periods, positions and launch profiles.

Among the principles were tolerance levels for psycho-motor performance decrements and visual gray-out, black-out and unconsciousness. Large individual differences were discovered in the ability of astronauts to perform piloting tasks in acceleration tests. Many of these principles were applied in eight Mercury flight simulation projects on the human centrifuge.

In the spring of 1960—before Alan Shepard's historic flight in May, 1961—a science editor from one of the television networks had arranged with the PR office to film a centrifuge run, in order to explain acceleration stress to the viewers. It turned out to be a monumental effort. Both the network's camera and the scientists' camera stopped functioning at high-G so in order to film the subject's face they had to solve that problem. The television crew and the centrifuge crew began working on this problem the evening before the program was to be aired live. They worked feverishly all night on the camera problem and other dilemmas, glitches and failures. Of course, I was not surprised that Randy joined them, but he had assured me that he was going to be a test conductor, not the subject.

When the time came to air the program, however, there was Randy in the centrifuge gondola—whether by design or default, I never knew. But what I will always remember is that the ensuing program which showed him whirling by, face contorted and heart alternately racing, thumping and then almost disappearing from the EKG, was the longest half-hour television program I have ever seen. It was even more traumatic for his mother who tuned in to see her son's TV debut from her Indiana home. His unsuspecting mother was unacquainted with the human centrifuge and ever-hopeful that her only son would settle down on some quiet university campus.

Instead, there he was flying through a 15-G profile with speech slurred and heart beating wildly. His mother, as well as the other relatives gathered together, nearly went into shock. These were the days before there was such things as "reality TV." However, the live centrifuge ride, which showed so graphically the stresses and strains of 15-G runs was about as "real" as anyone would want to see. The laboratory was flooded with members of the press and other concerned viewers who wanted to know how the subject was doing. It was never revealed whether this program was good publicity for the space projects or whether some of the aspiring astronauts gave up the idea entirely after watching it.

Most people—and I number myself among them—would not have volunteered for any of this, no matter how vital to the manned space effort.

We prefer to remain at a comfortable 1-G and to keep our eyeballs where they belong.

Navy Captain Kirk Smith, at right, Medical Director of the Aviation Medical Acceleration Laboratory, greets television crew members prior to CBS's historic broadcast of a centrifuge ride in 1959.

Early in the centrifuge studies, there was a group of scientists who regarded landing on the moon as a mere way station along the way to interplanetary travel, particularly to Mars.

In America's pioneer days, the prairies were dotted with way stations to provide creature comforts for travelers passing by on the Pony Express. Some of the space pioneers envisioned a similar role for a moon substation.

The Mars-or-Bust pioneers were seemingly undaunted by the obstacles which were still unsolved in the 1950s and 60s. They assumed that the moon landing would become a reality and they were simply looking beyond to the limitless possibilities of exploring outer space. Primary among the unanswered questions was human survivability. Were humans doomed to spend their lives in earthly comfort, or could scientists develop portable, sustainable life-support systems which they could carry with them into the great beyond? In the late 1950s, Drs. Carl Clark and James Hardy, along with their co-workers, designed an experiment to find some of these answers. The giant human centrifuge, which was being used to study gravitational stress and to prepare astronauts to survive it, was refitted for a Mars simulation. The largest hurdle which stood in the way of enabling a human to go to Mars is the time factor. Even when Mars is at its closest, it would take 9 months to travel there. A 9-month period in which the astronaut would require food, water, sleep, breathing air, normal atmospheric pressure, waste management, a fail-proof life support system, and a communication system not yet developed is no small order.

When these problems are all listed, they make sending robots to Mars appear to be the only feasible plan. But that was before Dr. Clark and Dr. Hardy and co-workers developed their astonishing alternative plan. This plan involves a law of physics which most of us lay persons find unfathomable. Simply put it is this: if acceleration stress is increased to 2-Gs—instead of the Earth's 1-G, velocity is squared every second. (That is, it would be multiplied by itself every second.) This phenomenon would be similar—in reverse—to a meteor which speeds through outer space until it is caught in Earth's gravitational field. Assuming that Mars was at its closest point and that a human could survive prolonged acceleration stress at 2-G and that he would maintain his pilot performance capabilities, the traveler could travel to Mars in hours instead of months.

Most people think they are looking at a typographical error. Mars in hours sounds highly improbable. Like most things that sound too good to be true, this fantastic trip poses obstacles never previously addressed. A human, turning himself into a meteor and rocketing through outer space, would challenge the science fiction writers, not to mention the space research scientists. The obstacles are numerous and complex.

Once again the irrefutable fact arises: humans are well adapted to living in the Earth's environment. The planet Earth is our home. However, it was not known whether a human could survive in such a hostile environment as outer space for prolonged periods at such velocity.

Sustaining 2 Gs for 24 hours presented challenges that had never before been addressed. It was not known if a human could continue to breathe under those conditions or if his heart would continue to function.

Scientists working on the life-support systems had all kinds of questions and concerns about what kind of protection the body would require. Food, water, waste management, breathing air, isolation, atmospheric pressure and earth separation were all factors which had never been tested before in a similar situation. There were other potential dangers lurking in outer space—especially for a long trip to Mars.

Earth's atmosphere protects humans from nuclear radiation belts. Nuclear radiation which exists in space poses a danger to the human explorer. Radiation sickness is fatal when fully developed.

In addition to radiation sickness in humans, nuclear radiation also cripples equipment, including the spacecraft. The design engineers were working to provide "radiation hardness" to protect the equipment.

Some of these obstacles were addressed by the AMAL staff with a special device called the G-capsule. This capsule was made of iron and filled with water or other protective fluids and was designed to protect the astronaut from nuclear radiation. Two pioneer space researchers, Dr. Carl Clark and Flanagan Grey, worked to develop the G-capsule. Flanagan volunteered to be the subject of a special test of the capsule. Encased in the G-capsule, the researcher survived a centrifuge ride of 32-Gs for 5 seconds, a record which is probably still standing. He was also well-protected against nuclear radiation belts. Unfortunately the G-capsule was too heavy to be considered practical for space travel.

At that time the Navy and the Air Force had developed nuclear airplanes. They were eventually cancelled because of the nuclear hazards.

It was hazards and obstacles such as these which made robots seem like a better choice for Mars exploration.

Two Gs—instead of the Earth's normal 1 G—may not seem like much, but actually it puts considerable stress on the human body. In this prolonged 2-G ride, the astronaut would go faster and faster as the bonds of Earth's gravity grew weaker. Along with this, the gravitational forces of other planets would also pull the vehicle along in what scientists call gravity assistance. At a certain point, the vehicle would rocket through space with a minimum fuel usage.

The day finally came when the 24-hour experiment was finally approved. Its purpose was to study the survivability of a human in such a venture. In research projects—especially the riskier ones—there are two factions: the risk-taker scientists and the government managers. The Mars project no doubt had its critics, but the project was finally approved. It will probably never be known what misgivings were held by the project planners.

The subject for this experiment was Dr. Carl Clark, one of Randy's research associates and well-known as a dedicated researcher. When I finally found out about this I was surprised (and relieved) that Randy wasn't the subject. The study proceeded as planned. The simulation was as close to the real thing as humans could make it.

The Aviation Medical Acceleration Laboratory was thoroughly prepared for the historic centrifuge run. The life support system, the medical monitoring equipment, drinking water, light, food, breathing air, and atmospheric pressure were all in place.

The subject also had equipment with which to test his motor skills, his visual acuity and his mental alertness. (These faculties were obviously very important for a space traveler who would have no one else to rely on as he traveled deeper and deeper into outer space.)

Manned space flight is much more than merely sending a human into space. The astronaut is also required to perform work such as operating controls and communicating with the home base. It is essential that the astronaut keep his faculties. He has to respond to visual stimuli, acoustical stimuli, instrument panel readings and timing and flight control devices. Sometimes acceleration stress causes gray-outs and black-outs or other visual disorientation such as illusions. Sometimes acceleration stress causes performance deterioration due to prolonged exposure.

A crucial question to be answered in this study was what the sensory thresholds were for each of the environmental stress factors.

On any flight, someone has to continually scan the instrument panel on the lookout for the first signs of malfunction of any system, especially guidance and control. Like more mundane mechanical things, when problems are detected early the chances of averting total failure are greatly improved.

Great attention was paid to these aspects because the mere idea of an unconscious astronaut hurtling through space beyond the assistance of ground control was too terrible to contemplate.

Although the word "caution" was not usually applied to research projects generated by Drs. Clark and Chambers, they proceeded cautiously with this study. Shorter duration centrifuge rides at higher Gs had been tried. The 4-hour test at 4 Gs had to be stopped because the subject could not take it. The 2-hour test at 4 Gs was stopped by the medical monitors when EKG monitoring showed acceleration loads for

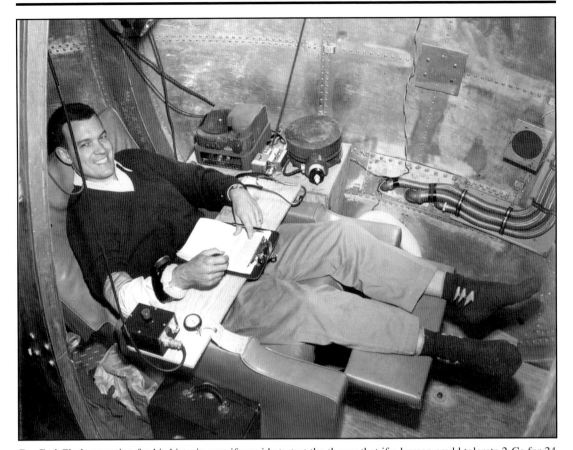

Dr. Carl Clark preparing for his historic centrifuge ride to test the theory that if a human could tolerate 2-Gs for 24 hours (instead of the Earth's 1-G environment) he could reach Mars in that time period. Dr. Clark concluded that travel time – if accomplished in a straight line – would take 30 hours because of deceleration time.

heart, blood pressure, respiratory, visual or other biomedical measurements had exceeded physiological tolerance limits. Even earlier experiments at 2-Gs for shorter periods presented some problems.

The engineers and scientists were eager to conduct the study, but the medical and safety specialists had great reservations and concerns. They insisted on continuous medical monitoring. As a precaution against medical emergencies, they had cardiac resuscitation equipment near at hand. The centrifuge gondola was also equipped with a "chicken switch" which could bring immediate medical assistance.

The 24-hour, 2-G run was to be conducted only once, unlike earlier astronaut training runs which were duplicated for many hours. This was unusual for a scientific experiment because scientists value research projects that can be duplicated and studied further. The preliminary studies of shorter runs had been meticulously conducted.

Earthlings function fine at the normal 1-G gravitational stress. Two Gs does not sound very daunting. Astronaut training runs were conducted at much higher Gs, but for much shorter runs. In a 2-G run for 24 hours, the subject would have to endure the pull of gravity at twice his weight for a length of time that had never been done before. The ability to sustain any of these conditions was essential to any manned trip to Mars.

Dr. Clark was monitored by medical researchers and flight surgeons for the 24-hour run. At the end of it, he was still able to talk and to move, although he reported being extremely fatigued.

And so it was demonstrated that a human could survive a prolonged centrifuge ride at 2-Gs which could take him to approach near Mars.

This was a significant first step—but there were many more hurdles to overcome. The robots which have landed on Mars landed roughly after bouncing numerous times. The 2-G study on human survivability was concerned only with getting to Mars. Preparing a human for a crash landing was another study.

Then, of course, there was the question of how to get off the planet Mars and head for home. There were no doubt many adventuresome souls who would volunteer to travel to Mars and return; there was practically no one who wanted a one-way ticket—or a franchise for the first Mars Bed and Breakfast.

Randall Chambers had hoped to be the subject for a second 2-G 24-hour test but it was never authorized. The centrifuge was refitted for research on the lunar projects and the focus was back on landing on the moon. The study had been a one-of-a-kind experiment and no Mars missions were planned at that time. So the completed study was put in the files and largely forgotten for the next 40 years.

Then in 1999 Dr. Chambers presented parts of the government report on the subject of human factors in Mars exploration at the annual meeting of the Mars Society. It was received with interest and astonishment.

Then Randy began receiving requests from young graduate students who wanted to study the Mars mission further.

In 2004 President George W. Bush announced plans to send humans to Mars. And the entire Mars-or-Bust enthusiasm resurfaced. Dr. Clark demonstrated that he could tolerate long duration 2-G in the centrifuge for 24 hours. He concluded that a human could tolerate a continuous 2-G flight in a near-approach to Mars, accelerating half way, and decelerating half way. He wishes he could have ridden the AMAL centrifuge for 30 hours. This may have been enough to get to Mars on its straight line near-approach from Earth. Drs. Clark and Chambers have emphasized that the challenge is to develop engines for a real flight that are capable of providing this acceleration.

For the engineers and scientists, more research is needed! And now—more than 40 years later—a new generation of scientists, buoyed by an array of high-tech equipment and youthful enthusiasm and energy, will work to figure out how to make a safe Mars landing and to return to Earth.

These are the same questions asked in the 1950s about the moon landing. Only the boundaries have been moved.

A description of what happens to an astronaut when he is launched into space orbit would never be included in a travel brochure. He first encounters tremendous acceleration stress as the spacecraft breaks the gravitational bonds of Earth. He then accelerates to an enormous speed and endures temperature fluctuations of heat and cold never encountered before by humans. And before he has time to recover from these assaults on his body, he finds himself in the strange environment of weightlessness (which scientifically speaking, is microgravity).

In scenes from science fiction, the astronaut floats around in apparent effortless glee and weightlessness appears to be a condition which all earthlings would envy. Free as a bird. The stark reality of weightlessness is far different. In the weightless environment the forces of gravity which hold everything upright and in place here on Earth are missing. So bodies and objects float around like feathers or balloons.

This means that the astronaut would be unable to stand up or to walk around or to sit down in the usual way. Instead, he would float around uncontrollably and maneuver himself along by grabbing onto things in the space cabin.

He would not be able to work with his hands in the usual way. For example, if he tried to turn a knob, it would be he, not the knob, that would move. If he were to be an effective pilot, he had to be able to press switches, operate levers, maintain his perspective concerning which way was up, and to be alert and adept enough to pilot the vehicle in an emergency.

The medical scientists were most concerned about the effects of prolonged weightlessness on the body. Without gravity, the heart would not have to work as hard. This would ultimately result in a weakening of the heart muscle with its resulting debilitation. Vital neurological and cognitive functions would also be affected.

Weightlessness would also weaken the body in numerous other ways—everything from atrophy of the muscles, loss of calcium and bone strength, to altering the healthy balance of blood chemistry and metabolism.

As with acceleration stress which was simulated on the human centrifuge, it was decided that a simulator would be the way to study weightlessness.

There were three possibilities. One was parabolic flights on jet airplanes in which the plane would climb to high altitude, then make a steep dive at high speed, creating a few seconds of weightlessness. The second choice was the use of a C-135 transport plane, which created up to 36 seconds of weightlessness at one trial during a steep dive at high altitude in a high speed maneuver. Both of these enabled scientists to study transitional effects from weightlessness to gravitational loads.

The air bearing simulator was also tested for moving objects in space since it minimized friction.

However, water tanks provided longer periods of weightlessness which were more valuable in these studies. (Water provided levels of neutral buoyancy, not genuine weightlessness, but it was decided that this would be close enough to give useful data.) The goal of the water tank experiments was that by the time the astronaut went into space he would be able to maintain his health and also feel at home enough to perform flight tasks. Water tanks ranged in size from one that was about the size of a large home spa, to the size of a deep swimming pool. (The difference between the water tank and a swimming pool, however, was that the water tank was 30 feet deep.)

The work progressed in painstaking steps. The first step was the development of special equipment and protective clothing. Among these were a variety of water suits, breathing apparatus that enabled the subject to breathe under water, water boots, gloves, helmets, visors and goggles. These sound like ordinary, everyday items. However, none of those in existence could have stood up to such harsh conditions and stringent requirements as needed for long-term missions. Each item was researched and prototypes produced. Then there was more testing and models were developed, followed by more testing, and then evaluation. Some of this equipment later was adapted for the life-support system for actual space flight.

The Navy's Sea Lab research projects also contributed and utilized the equipment. The initial water tank work involved placing the subject in water at neck level for periods ranging from 8 hours to 24 hours or longer. For the next experiment the subject was totally immersed in water—4 to 5 feet below the surface—to study his ability to reach out, press buttons, turn knobs, operate control sticks, maintain orientation and equilibrium, frequently while blindfolded and suspended. The water tank work was complicated by the fact that there were side effects on arm, hand and foot performance during prolonged submersion in water. Because of the skin maceration caused by prolonged exposure to water, one experiment was done in silicone oil. The subject was submerged in a large bag of silicone oil for 5 days, probably feeling somewhat like a canned sardine. There were no medical problems with the silicone oil but the experimenters went back to using water because of the high cost of the oil.

There were performance limitations in both environments. Water tank experimentation was a very dangerous situation for the subject. He was tightly fastened to a couch, chair or tilt table at the level of neutral buoyancy usually about 6 feet under water and solely dependent upon the various monitors to keep him safe. If any of the equipment had developed leaks—or if the monitors had become distracted—he could have drowned quickly, suffocated, or suffered injury. To safeguard the subject as much as possible, he had company in the tank: two certified Navy scuba divers. He was also continuously watched from the outside by a medical monitor and a data taker. Randy was so dedicated to this work that he took the Navy scuba diving course and became a certified diver. (As usual, I was the last to know.)

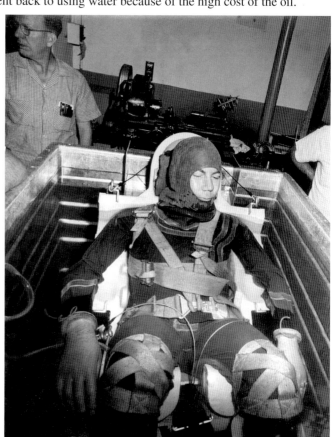

Another water tank study was an experiment in which the subject was submerged in a large wooden tank nicknamed the pickle barrel because it had been used to store large quantities of pickles for Navy food galleys. The pickle barrel experiment was as complicated as pickles are simple. The subject was placed on a tilt table

Submerged in his couch to the neck level in a water tank to simulate microgravity and neutral buoyancy, Lieutenant Colonel William Douglas, NASA Flight Surgeon, tries resting for long periods of time in this very early microgravity simulation. Richard Timmings, AMAL mechanic, observes.

6 feet under the water's surface. He was strapped to the tilt table in complete gear—breathing device, water suit, biomedical paraphernalia, helmet, footwear and gloves. Just getting him into the water with all of this gear was quite a challenge. They finally got him completely dressed and fastened to the tilt table. Then the entire unit was hoisted into the water tank. The tilt table was placed at different angles and the subject, who was blindfolded, was asked to point upwards. Much useful data was accumulated by measuring his perceived angles compared with the actual ones. This work was invaluable in understanding orientation—or disorientation—encountered in the weightless environment. Sometimes two subjects were placed on either end of the tilt table and their performances compared. They looked like two children lying down on a teeter totter.

(Left) The health of the subjects in these experiments was carefully monitored. In this picture monitors are checking Dr. Gordon's respiration and other vital signs during a water tank experiment.

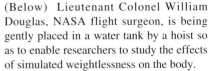

(Below) Lieutenant Colonel William Douglas, NASA flight surgeon, is being gently placed in a water tank by a hoist so as to enable researchers to study the effects of simulated weightlessness on the body.

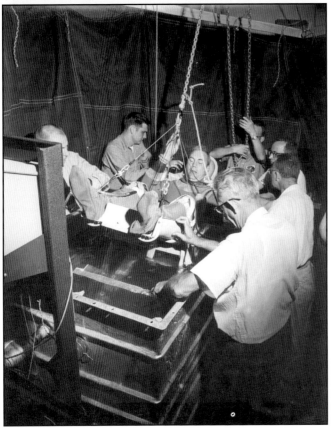

Another valuable study conducted in the large water tank at NASA Langley was the effects of microgravity on the body for long periods of time. To maintain a position approximately 6 feet under the surface, the subject was weighted down with lead weights sewn into his pressure suit. The subject practiced transferring cargo like some strange underwater mover. The effects of neutral buoyancy on the body and the ability to work in weightlessness, especially for long periods of time, were important for preparing for rendezvous and space station activity. Another study in the initial phase of water tank experiments was testing the entire unit on the centrifuge. The subject, in his complete attire, was strapped to a contour couch and placed in a water tank. Then the tank was hoisted onto a gondola hanging from the bottom of the centrifuge.

In this way the scientists were able to study the effects of sudden transition from 1-G to high G to weightlessness to high G, reentry, splashdown and back to Earth at 1-G.

Problems created or encountered in the weightless environment were studied, tested and discussed in the water tank research series. Eating—which is important to most of us—was a complicated procedure in weightlessness. After much trial and error research, the scientists came up with a plan and then a system. Basic nutrition in a form similar to pudding was placed in a squeeze tube. The tube was then connected to a pressurized tube built into the astronaut's helmet. The same type of fittings could be used to transport drinking water or breathing air—similar to the way the attachments of a vacuum cleaner can be changed. Sometimes the test subject connected the tube; sometimes, under water, an assistant made the attachment.

In later flights, such as those of the shuttle, food was placed in a closed microwavable container and then

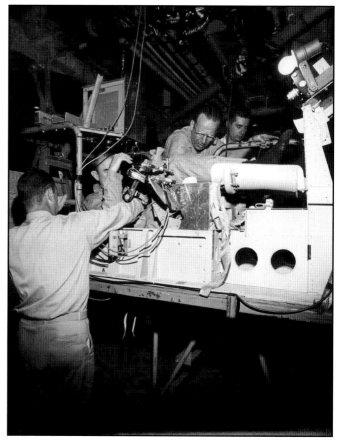

The subject – barely visible – is shown during one of the complex water-tank-to-centrifuge experiments.

eaten through a squeezable tube. This system was developed after a great deal of testing, research and trial and consultations with nutritionists. In subsequent studies, the food preferences of the subjects were considered. However, when a favorite food was liquefied to a consistency needed for the tube, it sometimes was altered beyond recognition. Food preferences and eating habits were very individual and efforts were made to consider these differences—within the enormous constrictions of weightlessness.

How to transport drinking water and store it was even more challenging. Drinking water was always at a premium and was strictly conserved. Sometimes, the water mixed with food was considered sufficient. There were also efforts to reuse drinking water by treating urine, but none of these alternatives was popular.

And then there was the subject which dogged scientists and astronauts alike from the beginning of manned space research: waste management. After much research about waste management, a system was developed. This consisted of bags that were placed inside the pressure suit and attached to the body. These were very uncomfortable, especially after use, and elicited many complaints. In the early days of space research, waste products were carefully saved and analyzed by biochemists in an effort to find a diet which would be most thoroughly consumed and produce the least waste. There were also efforts to develop drugs to suspend or enhance the digestive process or to suppress appetite. Excitement of the space adventure also complicated the picture. Some astronauts got so enchanted by views from the spacecraft that they neglected to eat or were unable to sleep. This caused serious consequences later in their physiological and psychological functioning.

Waste management has continued to plague the space traveler and tests their dedication to leaving the planet in the first place. In the beginning it was discovered that some people could chew and swallow under water and some could not without a great deal of practice. Digestive problems also presented themselves. The litany of symptoms sounded like those featured on television in ads of highly advertised remedies—stomach awareness, heartburn, unwanted gas, nausea, dizziness, vomiting and motion sickness. Eating and drinking in space had to be interfaced with other factors such as acceleration stress, weightlessness, exercise, sleep, and mission tasks. Mission performance was the objective of all flights and getting off schedule for any of the factors such as personal hygiene, eating, sleeping, communicating or photography upset the finely-tuned balance. The situation could be compared to preparation for surgery when the patient is carefully prepared and monitored before and after the procedure.

Much effort was made to develop a first aid kit to combat and alleviate some of the health problems. However, the astronauts were so different from each other and had such diverse problems that this was almost impossible.

As this work progressed, it became increasingly clear to earthbound people that an astronaut would have to be truly dedicated to tolerate the discomforts of space travel, no matter how exciting the adventure might seem.

In addition to preparing for physical survival in space, the water tank experiments also involved practicing work-related skills. Such skills as cargo handling, transfer and storage were practiced in the water tanks. Practice was conducted for both EVA (extravehicular) and IVS (intravehicular) conditions. Transporting food, water, containers of supplies, equipment and tools in weightlessness was greatly complicated. Astronauts had to learn to walk and carry cargo to specific locations in the water tank and place them in compartments for storage and later use.

Sometimes the practice was conducted by submerging two spacecraft in the water tank and practicing transferring cargo and tools from one to the other under water. Sometimes tether lines were used to further practice precise placement of objects. This was a serious study designed to prevent an essential tool or cargo from floating away from an astronaut in space. Tether lines were also used to attach the astronaut to the spacecraft during these practice maneuvers. The challenge facing the astronauts in preparing for this task might be compared to loading supplies and equipment for a cruise ship in preparation for a week at sea. Except in outer space, the task would be greatly complicated by the weightless environment. Skills practiced in the early water tank experiments were one of the life-saving factors in the rescue of Apollo 13. Those astronauts called upon everything they had learned and practiced in simulation studies. Orientation—without the usual visual, vestibular and other sensory cues—was also studied extensively in water tanks. At first, staying upright in the weightless environment required a tether at the waist and feet. After training, longer tethers could be used to expand the work area. The water tank experiments were used not only to train astronauts but also to design their equipment, including pressure suits.

In addition to preparing an astronaut to work and survive in the weightless environment, the water tank studies also made important contributions to the field of medicine and psychology. There is a physiological phenomenon called orthostatic hypotension that results from sitting or lying down for a prolonged period, and then getting up suddenly. Your blood pressure drops and you can get dizzy, feel faint and nauseous, and may pass out. This was potentially a very real problem for astronauts spending long periods of inactivity lying in a bent safety pin position in a space compartment in the weightless state. A way of simulating this condition was through prolonged bed rest studies. One of the early—and most memorable—studies of this problem was done by a young physician, Dr. Cleto DiGiovanni, who was a U.S. Public Health Service Fellow who worked for Randy. Cleto kept insisting that orthostatic hypotension needed to be addressed early on. Otherwise, he warned, an astronaut could have serious consequences if he became upright suddenly after lying for a long time in a cramped space such as a space capsule. The Navy, however, viewed other issues as more pressing and believed that space suits would help protect the astronaut from this problem.

Cleto was a maverick, prone to act without authorization if the cause warranted it. Randy was his immediate supervisor, which was not a good arrangement because Randy is the same type. To make his point, Cleto, one Saturday night, packed up a space suit and a contour couch and decided to demonstrate this phenomenon in a Navy hospital. He recruited some corpsmen to help him. The corpsmen were well-trained and conscientious, but the scene before them was a challenge. Space suits were a rare sight in those days, contour couches even more exotic. And this "patient" was lying in a bent-safety pin position and refused to take off his boots. Word soon spread that there was an astronaut from outer space lying in one of the rooms in full space gear. Around midnight, someone called the hospital administrator with a report of a space alien in his hospital. Understandably, the administrator, who probably thought he had already seen everything, came down to see this space man for himself. He called Randy at 5 o'clock the next morning to ask him if he had a medical student working for him. He thought that Cleto was engaging in a medical fraternity initiation rite. Cleto lay in the bent safety pin position in the "protective" space suit for eight hours. When he got up, he passed out cold and required real medical attention.

This pilot, dressed in space gear and reclining in a contour couch in a "bent safety pin" position, appears similar to the way Dr. Cleto DiGiovanni must have looked that Saturday night at the Navy hospital.

He and Randy later reviewed what was known about the physiological and psychological aspects of the gravity spectrum in a seminal three-part publication in the prestigious *New England Journal of Medicine*. Eventually, space physicians began dealing with these physiological problems by incorporating exercise into the astronaut's schedule in flights such as those of the shuttle and the space station.

As the water tank work progressed the scientists decided they needed to study performance skills more thoroughly and for that they needed a larger tank. They set their sights on a Philadelphia aquarium which was the home of a group of dolphins who regularly entertained visitors. The scientists built a barricade to confine the dolphins to one end of the large water tank and then proceeded to conduct performance studies in the rest of the tank. I didn't find out about this project until after the fact, so I was spared the embarrassment of watching the aquarium visitors reacting to the impromptu program. They were no doubt amazed to see scientists walking a plank blindfolded at one end with puzzled dolphins at the other in a strange double feature. I never found out if they were disappointed.

The aquarium was subsequently used for fatigue studies also. Astronauts traveling into space are required to stay alert and awake for long hours so this had to be part of their preparation. The water tank research has had a wide-ranging effect on medical treament. The studies on weightlessness cast serious doubts on the wisdom of prolonged bed rest which had been a standard remedy for centuries. In modern hospitals, surgical patients are placed on their feet shortly after surgery. Physical therapists are sent to the bedside of most non-infectious patients. Practically no one is allowed to "enjoy" prolonged bed rest any more. This change is the direct result of the weightless studies done for astronauts in preparation for space travel. The water tank research—at both the Navy lab and the public aquarium—was done with the scientists' usual disregard for what people would think. This was apparent in an incident that happened on our doorstep in the midst of the studies.

In those days, milk was delivered to the door several times a week. Early one morning Randy came home from having spent the night in a tank of water as a subject. He was very clean and he had a shriveled up look about him.

He arrived home just as the milkman was making his delivery. For some reason, Randy decided to give the milkman an explanation of his scrubbed appearance, evidently unaware that his explanation would only create new questions.

"I've just spent the night in a tank of water," Randy told the milkman. At first, the milkman looked skeptical. Then he gave Randy a look which said that this was a really neat alibi that all married men would envy. As unlikely as it appears to the non-scientist, the water tank simulations were proved to be essential in preparing astronauts to survive and work in microgravity. As space research proceeded, as soon as they reached one milestone, they began planning for the next. They still had a long way to go.

The five biggest challenges to preparing humans to travel into space and return safely are acceleration forces, weightlessness (microgravity), atmospheric pressure, life support and pilot performance.

Gravitational stresses have to be conquered in order to survive launch and reentry. The weightless environment must be understood to enable an astronaut to survive and function in outer space. The atmospheric pressure factor inserts itself into both of these stages and then some. This problem can be illustrated by a pet store purchase. When we buy a gold fish, it is placed in a plastic bag of water for the trip home. Water is the environment in which the fish functions and he would soon perish without it.

Similarly, astronauts trying to travel into outer space have to take their atmospheric pressure with them. How to package it and transport it—in a fail-safe way—was a challenge which loomed large over the early space research projects. Pilot performance, not merely to survive but also to work, required the coordination of all of these other factors.

Atmospheric pressure is among the least understood of all of the obstacles to getting off the planet. To most of us, an astronaut with his pressure suit, his portable oxygen tank and his spacecraft appears to be all set to travel into space. In reality, atmospheric pressure, taken for granted on this planet and mostly absent from outer space environments, complicates a journey into space.

The medical profession calls breathing a "vital sign" for a reason. The astronaut could not go very far nor stay very long, without breathing air provided by the atmospheric pressure on his home planet. Airline passengers experience a little introduction to this phenomenon when the stewardess announces that masks will descend from the compartment above each seat to provide breathing air if the cabin begins to lose pressure. Even though airplanes do not ascend into the stratosphere, airplane cabins are pressurized because loss of pressure would soon render passengers unconscious.

There's more. Atmospheric pressure not only is crucial to breathing, it affects the space traveler's performance as well as his clothing, equipment, and all the materials that go into them. Atmospheric pressure and gravity both decrease with altitude and orbital flight and together create a sort of double whammy to the physical and mental functioning of humans and the performance of their equipment in outer space.

Atmospheric pressure is a combination of life-supporting gasses and gravitational forces and plays a crucial role in human physiology. The forces of gravity continually help move the blood throughout the body. The heart then reacts to gravity and pumps blood to the body's extremities to complete the circulatory system.

Gravity also has a role in breathing, forcing the lungs to contract and then to be expanded again. Oxygen must be available at sufficient partial pressure in the breathing air for adequate respiration.

But it's even more complicated than that. Atmospheric pressure affects various other pressures such as cabin pressure, lung pressure, intrapulmonic pressure (within the lungs), blood pressure, oxygen partial pressure, arterial pressure, and alveolar oxygen pressure (which assists the body in absorbing the oxygen). Proper ratios among the cabin pressure and bodily pressures must also be maintained. Otherwise, decreases in pilot performance, vision and other senses, and even loss of consciousness would result.

Unless adequate atmospheric pressure is maintained, the astronaut's clothing and equipment would not hold together either. For example, a bottle of water will burst if it is subjected to unbalanced pressure between the inside and outside of the bottle. Other equipment might melt, fall apart, disintegrate or otherwise fail if not made of special materials developed to withstand the harsh conditions.

As if all of this wasn't complicated enough, temperature and humidity, and individual differences also were part of the problem. In simple terms, atmospheric pressure is a many-faceted factor which presented a big headache to early life scientists and physiological researchers. Human performance requirements in outer space are much more stringent than earthly ones. A flight into outer space leaves little room for human error or equipment failure. Here on Earth, equipment failure may be inconvenient but is eventually resolved. The failure to control atmospheric pressure in a spacecraft or any of its equipment could be catastrophic. The air which humans are accustomed to breathing is heavy, weighing 15 to 16 pounds per square inch depending upon the altitude at which it is measured. It also contains a finely-balanced combination of partial pressures of oxygen, nitrogen, water vapor and carbon dioxide.

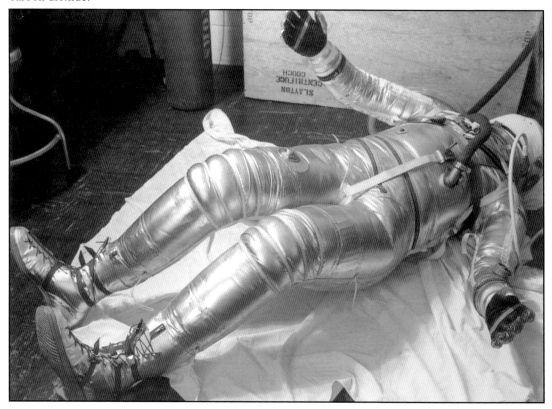

The Man Who Wasn't There. This is a picture of an early Mercury pressure suit inflated (empty) so that technicians can check the suit for pressure levels and leaks.

One reason air is so heavy is the force of gravity which strongly pulls everything toward earth. That's why the higher one ascends, the lighter the air becomes. At less than 3 pounds per square inch of oxygen partial pressure a person is likely to lose consciousness. The minimum requirement is at least 3 pounds per square inch at 74 degrees. There must also be appropriate pressures of the gas mixture within the pressure suit, helmet and space cabin. In case of emergency the astronaut had to be prepared to manually adjust the controls for the breathing air.

Without sufficient total atmospheric pressure, vaporization would occur. On the other hand, too much pressurized oxygen would create oxygen toxicity and a fire hazard. These stringent parameters cast grave doubts that there could be life—as we define it—on other planets. To study the effects of atmospheric pressure, or its absence as in a vacuum, the scientists turned to atmospheric pressure simulators and testing devices.

The phenomenon was studied in a variety of pressure chambers both large and small. At the beginning of the space research, scientists utilized a major atmospheric pressure chamber located in a large room adjacent to the human centrifuge at the Aviation Medical Acceleration facility. It had been used to test flight personnel with and without pressure suits during the development of military aviation.

The chamber at the naval laboratory consisted of a large room which could be converted into a vacuum. There also was a larger atmospheric pressure chamber at the nearby naval base in Philadelphia. This was a large metal cylinder which was big enough to hold a crew of six.

As the space research progressed, the gondola of the human centrifuge was pressurized to add realism to the studies of acceleration stress.

Research into atmospheric pressure has been conducted for many years for the development of submarines. The breathing problems in submarines are the opposite of those in outer space. Deep in the ocean, the environment has much too much pressure because it is a combination of air pressure and the ocean's tremendous water pressure. The problem down deep is to reduce the atmospheric pressure to 7-to-15 pounds per square inch ranges. Most of this work was done at the Navy's submarine base in New London, Connecticut.

The problem in outer space is also to keep atmospheric pressure in the 15-pounds-per-square inch range, but working from an outside environment that has little or none. Research and testing to solve the problems of atmospheric pressure were done simultaneously with the early centrifuge studies.

Pressure chambers ranged from very small—the size of the astronaut's pressure suit or the centrifuge gondola—to very large pressurized chambers that were 40 by 60 feet or more. These were located at various Navy, NASA, Air Force and Army test facilities. The small ones were used to study human reactions; the large ones were used to study vehicles and equipment. Some were used to test the entire systems of pilot, equipment and vehicle.

In the early 1960s new atmospheric pressure simulators were being built at NASA Langley. These bigger atmospheric pressure chambers were big enough for the entire spacecraft or even huge water tanks for weightlessness studies. New wind tunnels to study aerodynamic effects on vehicles to prepare for launch and reentry and to give astronauts a chance to practice guiding the vehicle during launch and reentry were also built.

There is an old saying that a system is only as strong as its weakest link. Space researchers, with their meticulous attention to detail, studied atmospheric pressure conditions on the performance of the system, the equipment, the human operator in the atmospheric environment and the pilot's ability to pull all of these parts together.

A major part of the atmospheric pressure training was for emergency situations. If a leak should occur in any part of the system, the astronaut had to be prepared to save his own life and those of the crew. He was presented with a series of simulated malfunctions to which he had to respond appropriately by operating controls and displays. Redundancy and safety switches were built into the systems to buy time for an astronaut facing an emergency. Since the astronaut was going to be alone in outer space, there would be no one close enough nor fast enough to rescue him.

One example of this emergency-type research was the case of astronaut John Young simulating reentry on the centrifuge at 14-G for 5 seconds. Even though he lost his vision, he could operate the control stick and manage the spacecraft flight profile by touch and memory. In John Glenn's Friendship 7 flight, he had to call upon his emergency practice sessions to manually manage the reentry when the heat shield retro-pack malfunctioned. Years later, in the Apollo 13 emergency, some of the simulated centrifuge emergencies were employed for real.

Human beings are in many ways the weakest part of the man-machine space system. This is true, in part, because equipment can usually tolerate more stress than humans. However, human beings are—without doubt—the best problem solvers.

One fact, which would send chills through the faint-hearted, is that even the most carefully-designed emergency features in the space system occasionally malfunctioned. An emergency feature on the blink? All alone in the great beyond? This would really cause many of us earthlings to wring our hands in dismay. But astronauts were trained and prepared to take over manually for malfunctioning systems.

Another contributing factor is the sheer complexity of space travel. As the spacecraft went through essential events, the temperature, gravitational stresses and the atmospheric pressure fluctuations combined to present a series of rapidly-changing conditions. Each of the following essential events presented different profiles which had to be monitored and controlled. Responding to them in appropriate sequence is enough to make your head swim.

Examples of events practiced in simulators occurred in the following time sequence: escape tower jettison, capsule separation, retro sequencing and initiated, retro attitude confirmed (or denied) by telemetry, retrofire, retro pack jettison, retract scope, snorkel, drogue chute, and then main chute. The first 6 stages involve lifting off and entering Earth orbit; the last 4 involve reentry through the Earth's atmosphere and splash down. Space travel is not for the faint-hearted, the absent minded nor the slow-responders!

Some of the earliest experiments were performed by pressurizing or depressurizing the space cabin on the human centrifuge while the astronauts were being trained to accommodate to acceleration stress. This gave the astronaut—and his trainers—a picture of the real adjustments to come. The transition to these combined effects—acceleration stresses, severe temperature fluctuations, and weightlessness all within an envelope of transported atmospheric pressure has a complexity hard to comprehend. Maintaining appropriate atmospheric pressure levels for the astronaut was so crucial that the astronaut's physiological and psychological performance were measured at each level along the flight profile.

Equipment adjustments and modifications were performed on the pressure suit and all of the equipment for each phase of flight. In addition to the outside monitors of each aspect of the practice flight, the pilot also made a verbal report of his observations and opinions. Sometimes the routine reporting was enlivened by a complaint or an especially low score the astronaut gave a procedure on his rating scale with added comments.

A NASA researcher tests the inflated pressure suit, checking for leaks.

In the early days of space research, the scientists started with the basics. The cone-shaped capsules of Mercury, Gemini and Apollo were compared to the glide vehicles of the X-20 and the space shuttle. Astronaut training for each of these different designs was also very different.

Other basic questions were also asked and answered. Should the astronaut face the wind or turn his back on it? What were the best positions in the spacecraft for the pilot to perform at his best? These questions—and many others—were answered after much initial research and testing.

The Mercury Project astronauts were tested, examined, rehearsed and outfitted, receiving more attention than most other human beings in history. In the Mercury studies the astronaut was strapped into the contour couch, custom made, and wearing his specially developed pressure suit, helmet, gloves, boots and biomedical instrumentation. He was presented with his flight control stick and cockpit instrumentation and challenged with the cockpit display panel to fly a Mercury mission while exposed to atmospheric pressure changes in the gondola. He also wore a microphone and ear phones, as well as temperature and respiration sensors inside his helmet for vital feedback with the medical project personnel.

His pressure suit had been carefully, meticulously developed to protect him from the atmospheric pressure changes. Acceleration of launch and reentry in the Mercury profiles added realism by providing the entire package of acceleration stress, weightlessness and atmospheric pressure. Each phase of the flight was simulated as the astronaut performed the appropriate flight maneuvers. Atmospheric pressure within the cabin and atmospheric pressure conditions outside the cabin had to be monitored and controlled.

Later on, in larger pressure chambers the entire space cabin was studied with its equipment. The performance of the astronaut, as well as the performance of his clothing, equipment and the spacecraft itself, had to be studied in various atmospheric pressure environments.

All of these early studies were the prelude to some of the most comprehensive, complex research studies in the annals of science.

The idea was to combine all of the elements which had been studied separately—gravitational stress, weightlessness and atmospheric pressure—along with the astronaut performance designs, the protective clothing, the equipment, the biomedical monitoring—to create the space flight conditions and outer space environment here on planet Earth. The research series was both frustrating and exhilarating to the scientists.

The idea was very exciting, especially since very few studies like this had ever been done before. The frustration came with the limitations which kept coming up. For example, the facilities were not big enough to contain the water tanks, the Mercury capsule and

Project Mercury astronauts Virgil Ivan Grissom, Captain, USAF of Mitchell, Indiana and Alan Bartlett Shepard, Jr., Lieutenant Commander, USN of East Derry, New Hampshire, shown comparing notes after being subjected to the Mercury-type acceleration and low pressure profiles in the gondola of the human centrifuge.

other paraphernalia at the same time. The closest analogy that comes to mind would be preparing for one of those reality television shows on a desert island and then bringing in supplies with a helicopter.

Scientists are among the most thorough and meticulous types in the human race. So they began their combination "reality" experiments with a bed rest study. This requires some explanation. Long-term bed rest affects the body in many of the same ways as prolonged confinement and microgravity in a spacecraft.

Astronauts were launched into space in a very small spacecraft and spent flight time in cramped positions in a microgravity environment. This resulted in the loss of vital chemicals needed by the body for normal functioning and the loss of calcium and strength in bones and muscles and also could affect vision, hearing, balance and other variables which in turn impair pilot performance.

The plan for the bed rest experiment was to take a group of healthy young volunteers—athletes from nearby colleges—and put them to bed in a Philadelphia hospital for 30 days. To be as realistic as possible, the subjects were not allowed to get out of bed at all. All studies involving human subjects have unexpected, unplanned events.

The college athletes who volunteered for this study soon discovered that 30 days of bed rest is not the easy life they had imagined. Several of the subjects became so restless that it soon became necessary to have monitors watching them to keep them in bed. And some subjects quit the study completely. With their athletic fitness diminishing and their boredom increasing, they evidently decided that the space program could continue without them. The subjects were not the only ones who began to see the punishing effects of long bed rest.

Flight surgeons, hospital medical staff, other NASA and Navy project directors all began to have concerns about the effects of prolonged bed rest. A study was also conducted with chimps as subjects and the debilitating results were similar.

Out of these concerns emerged the emphasis on physical therapy programs developed by modern medical facilities. Back in the 1950s and 60s patients were allowed to have bed rest after surgery or illness but the space bed rest studies persuaded the medical researchers to take another, more critical look at bed rest and its consequences.

In addition to confinement and inactivity, the astronaut also experiences microgravity, which has less pressure and further weakens the body. Microgravity also affects the brain. Since human performance is controlled by the brain, this has enormous consequences for astronauts. They are called upon to perform precise tracking tasks and other flight maneuvers in time-line sequences. To study the microgravity effects, the bed rest subjects were scheduled to be immersed in water tanks immediately after being removed from bed. This was supposed to simulate the sudden change experienced by the astronaut when he is launched into space and immediately plunged into weightlessness.

However, the scientists discovered that real time plays havoc with simulation. The hospital and the research facility were far apart and plagued by Philadelphia traffic jams which precluded real-time efforts. The scientists then tried to compensate for the time lag by having the subjects transported from bed to a water tank which was then loaded and transported by ambulance. The ambulance drivers were cautioned to drive smoothly rather than their usual emergency speed. (So far as is known, no one ever asked the ambulance drivers what they thought of these goings-on.)

The subjects, still in the water tank, were transported to the research lab and then suspended on the centrifuge and taken for an acceleration profile ride. They may have thought that they had fallen into the hands of some mad scientists at that point! Some of the bed-rest volunteers were placed in isolation chambers after the bed rest.

Various isolation tanks, rooms, chambers, and other facilities were used to create an environment in which the subjects were deprived of certain sensory stimuli. This was intended to study the effects of isolation experienced by a lone astronaut in space thousands of miles from other humans or civilization. The main problem in isolation was its effect on energy and activity levels which would have a serious effect on performance. Everything from respiratory activity, metabolism, energy requirements, cognitive skills and alertness, heat and sweating are affected by inactivity and isolation. Motor skills essential to walking and the neuromuscular skills needed for operating equipment were essential to the flight's success.

Performance problems were taken very seriously because when an astronaut was all alone in outer space his very survival depended upon it. These "combination" studies involving gravitational stress, weightlessness, atmospheric pressure, bed rest and isolation and bizarre performance tasks, such as cargo transfer in space, illustrated again the complexity of astronaut preparation and training. And the lengths to which the scientists went to create the hazardous environment of outer space. There were some humorous sidelights to the situation.

Three astronauts, John Glenn, Neil Armstrong, and Scott Carpenter agreed with the scientists that they needed every bit of help that was being offered. However, some of the others thought that some of the research projects were too harsh and demanding, that a lot of the meticulous research and testing was unnecessary. They frequently reminded the scientists that military pilots were experienced, competent and ready to go. What they failed to realize is that unlike other facilities they had experienced, they were in a medical research laboratory. They had fallen into the hands of flight surgeons, psychologists and physiologists. And all the while the astronauts were impatiently insisting that they were already prepared for space travel, the extensive biomedical monitoring told a different story.

Blood pressure, heart rate, respiration and voice response measurements indicated that they were overly excited or apprehensive. Occasionally the countdown was delayed to help them calm down. The scientists let them think that the delay was caused by equipment failure so as not to alarm them further. It could be said that no one could have done more to prepare them for exploring such a strange, dangerous, unfamiliar environment.

When the space program began, research projects on various aspects of the tremendous challenges proliferated at research and development centers in widespread parts of the country. Here is a sampling, with apologies to those who feel they should have been included, but were not mentioned.

Many centrifuge studies and other research projects were done to study different aspects of acceleration stress. At the Pensacola Naval Air Station U.S. Naval Aviation Medical Center in Florida scientists studied slow rotation room effects for humans confined in space stations as well as other confinement and isolation problems. At the USAF School of Aviation Medicine, Randolph AFB, Texas, medical effects of weightlessness were studied in aircraft and space cabin simulators. At Holloman AFB the effects of extremely high-G, impact and crash were researched and tested. At Wright-Patterson AFB Aeromedical Laboratory in Ohio the short-armed centrifuge was used for flight simulation and human factors studies. Scientists there also had an array of drop towers for high impact studies and a large human engineering laboratory which studied design and control of flight equipment.

Advances in biomedical monitoring were made at Wright Patterson as well as at Lovelace Foundation in New Mexico. The University of Pennsylvania Medical School and the Aerospace Medical Research Department of the U.S. Naval Air Development Center also developed and tested biomedical monitoring equipment. The Pennsylvania Naval Center, NASA Ames in California and NASA Langley in Virginia also conducted research and training on pilots and astronaut performance.

Confinement and isolation—daunting problems in long space flights—were studied intensively at the naval base in New London, Connecticut, which has a long history in submarine research of similar problems. NASA laboratories at Huntsville, Alabama, specialized in research and development of rockets and spacecraft. The NASA Lewis Research Center in Cleveland, Ohio, tackled the long list of safety questions. NASA Ames Research Center in California concentrated on human factors and life support research and development for all space systems. NASA Manned Spacecraft Center in Houston built a centrifuge for research and pre-flight refresher training. Also, water tanks and microgravity simulators were built for EVA studies and astronaut training. NASA Langley, which had the distinction of being the first space flight center, concentrated on advanced research and design. With their vast array of simulators they performed what might be called "second generation" testing and refining of basic research. Walter Reed Army Institute in Washington, DC concentrated on behavioral and medical research. Also, the U.S. Army Research Institute in Alexandria, Virginia, concentrated on behavioral research such as crew compatibility and team performance. The U.S. Army Research Institute at Natic, Massachusetts conducted on-going studies on food and dietary requirements for astronauts as well as the military services.

Aberdeen Proving Ground in Maryland and the NASA High Speed Test Flight Center in California joined the Natic group in developing and testing medical and behavioral health standards for space travel. NASA Goddard research laboratory in Baltimore, Maryland, did research in biology of living organisms in space. Goddard scientists were also involved in the search for life on other planets. The School of Aerospace Medicine and the Human Resources Research Center at Brooks AFB in San Antonio, Texas, studied the effects of acceleration stress and also tested life support systems. The Pensacola Naval Air Center in Florida studied disorientation and inner-ear disturbances in research in a rotating room. Subjects were placed in the room for several days. This data was needed to prepare for long space trips.

Lunar orbit, rendezvous, docking and lunar landing research and training was conducted at Houston Manned Spacecraft Center and at NASA Langley in preparation for the moon landing. Extensive training was conducted on the LEM and the Rover and the procedures for landing and exploring the moon. Many aerodynamic research projects were performed on the Apollo command and service modules and flight crews.

Research and training in flight dynamics was continued at contractor facilities in Houston, Cape Kennedy, Huntsville, and Los Angeles.

These research centers—and others— were all well-supported by many contractors from industry and university laboratories, both in this country and abroad. One of the first foreign research centers was the Defense Medical Research Laboratory in Toronto, Canada. This center was equipped with an array of flight simulators.

All of this research, testing and development culminated in what might be called "the longest dress rehearsal in history!" Shown here are more pictures depicting early training and research.

(Left) This close-up view of the AMAL human centrifuge gondola shows an astronaut at the controls. A vigilant monitor is in the overhead control station.

Many human centrifuge flight simulations of the Mercury acceleration environments were planned using this early Mercury-type instrument panel and cockpit controls installed in the centrifuge's gondola.

(Above) An exterior view of AMAL's modified human centrifuge at the U.S. Naval Air Development Center. The simulated Mercury astronaut life support system in the gondola can be seen at the end of the centrifuge arm. Thousands of hours of research and training were spent by scientists, pilots, astronauts and engineers studying and training for problems posed by the acceleration stresses of launch, reentry and gravity in partial-pressure.

(Right) Astronaut John Young enjoyed his training on the AMAL Human Centrifuge. Most of the astronauts enjoyed this machine. As a static and dynamic space flight simulator, it offered flight flexibility for training and testing on a large variety of spacecraft simulations and flight maneuvers. It was used as a "total mission" dynamic flight simulator.

(Left) Dr. Chambers records Astronaut Shepard's comments following a series of Mercury centrifuge familiarization runs.

(Below) This group of test pilots from the U.S. Air Force Aerospace Test Pilots' School seemed to enjoy their visit to the Naval Air Development Center and their rides on the human centrifuge at the Aviation Medical Acceleration Laboratory.

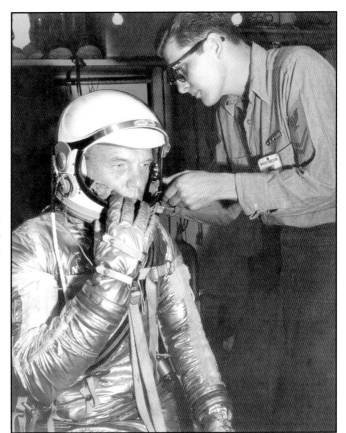

(Right and below) Astronaut John Glenn shown receiving final adjustments to his Project Mercury full-pressure suit prior to entering the gondola of the human centrifuge.

(Left)   Astronaut Alan Shepard enters the human centrifuge for refresher training in simulated G-forces, life support and pilot performance tasks.

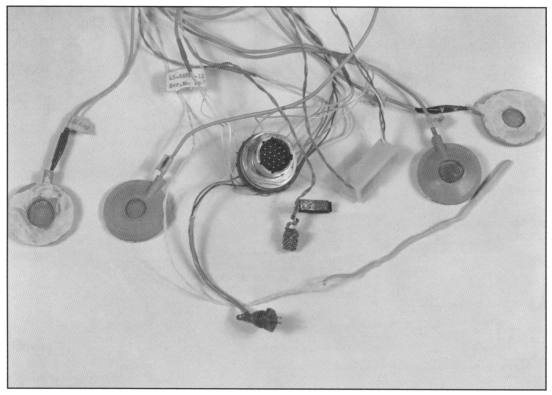

An example of the bioinstrumentation that the Mercury astronauts wore under their pressure suits. These instruments allowed researchers and medical staff to monitor the astronauts' vital signs during simulation testing.

Mercury astronaut Malcolm Scott Carpenter, Lieutenant, USN, shown receiving final adjustments to his Project Mercury full-pressure suit prior to entering the gondola of the human centrifuge.

Astronaut John Glenn, restrained in his contour couch, operates control devices to prepare for Mercury centrifuge simulation. His helmet is secured by a special locking device that was not used in actual flights.

A major effort in preparing astronauts to venture into space was developing the special protective clothing and equipment which was meticulously designed, manufactured and tested. That challenge was even more complicated than it seems and the clothing was more unusual than the casual observer might imagine. In fact, the outfits were a unique combination of garment and equipment.

(Included in this chapter are pictures of clothing and equipment taken at various stages of selection and testing.)

The first space researchers utilized G-suits which were borrowed from high flight aviators. (Some versions of the G-suit were actually used in World War II.)

The G-suit protected the aviator at high altitudes with either straps around the lower extremities or partial air pockets to help keep the blood from being forced out of the brain and pooling in the legs which would render the pilot unconscious. Although some G-suits are still used in aviation, it was soon discovered that this idea was inadequate for astronautical flight. Atmospheric pressure and composition which were such a major obstacle to space travel, loomed large over the clothing and equipment project. A major reason for this was that after each item was designed, created and tested, the entire ensemble—if it could be called that—had to be filled with breathing air.

A test pilot from NASA Ames is shown testing his G suit on the centrifuge in Pennsylvania. A careful look at his left arm shows a tube attached to monitor his blood pressure.

This process was not unlike inflating balloons for a child's birthday party. Most of the Mercury cabins and suits were pressurized with 100% oxygen at 5.1 pounds per square inch. This compares to the user-friendly atmosphere on Earth of approximately 20% oxygen and a delicate balance of other gasses such as nitrogen and carbon dioxide.

The 100% oxygen environment was selected for space travel because it was found to be more protective of the human body during high gravitational stresses of launch and reentry. It posed other problems such as a potential for fire. It also complicated both the design and the material for the clothing. Transporting breathing air is made even more difficult by the need for refilling the supply as it is used. The small suitcase-like container carried by the astronaut maintained the proper pressures and quantities in both the suit and the cabin.

The astronaut's suit and the cabin had to be reliably leak-proof, for the astronaut's survival depended on it. The astronaut could check his suit for leaks using two barostats inside the cabin. They did not pinpoint the location of a leak and there was no way to mend it anyway. Upon discovering a leak the astronaut or ground control could abort the mission if necessary. The pressure levels of the suit and the cabin could be adjusted according to the astronaut's requirements much like adjusting the thermostat in a room.

Similar challenges of clothing design could be found in earlier times in the design for deep sea divers and high altitude pilots. They, too, had to surround themselves with breathing air in order to survive the hostile environment to which they ventured. Space scientists took flight suits—which had in earlier times also been carefully researched and developed—and altered them to meet the extra requirements of the harsher conditions of outer space.

With the fervor of fashion designers, but with entirely different goals, the scientists worked on each item of clothing for this new venture. They began from the skin out with a very old-fashioned clothing item: long underwear. It looked very much like that worn in pioneer days in harsh environments. Droopy and unfashionable as it appears, long underwear was used as the first layer of astronaut's clothing because it allowed for ventilation. One can only imagine how uncomfortable clothing that is completely sealed up would soon become.

The first time I saw a test pilot in long underwear I wondered if they would find anybody willing to appear before the world's media in such unflattering clothing. (I thought this would surely rule out female astronauts especially!) Nevertheless, variations of old-fashioned long underwear were worn by all of the early space travelers.

Over the long underwear was the flight suit for lower altitudes in general aviation or the pressure suit for space flight. Various contractors sent different versions of pressure suits to the medical laboratory to be tested by scientists and astronaut candidates.

Flame retardation, comfort, flexibility, mobility, insulation to help the body adjust to wild temperature variation, and ease of putting it on and taking it off were major factors to be considered. (No one thought a pressure suit could be worn very long, so the plan was for the astronaut to take it off after he reached orbit. Then, of course, the suit had to be put back on to prepare for reentry.)

Clothing that was too complicated was ruled out. The suits had numerous zippers, long and short. They were at the neck, connecting the helmet; the wrists attached to the gloves; across the chest, around the ankles attaching the boots, with straps holding them into the contour couch. There also were connectors to which pressurized oxygen tubes were attached and up to 16 electrical leads for the biomedical monitoring, communication and pilot performance monitoring devices. The astronaut would be all alone in space and the thought of him caught in a tangled zipper or connector was another nightmare scenario for the scientists.

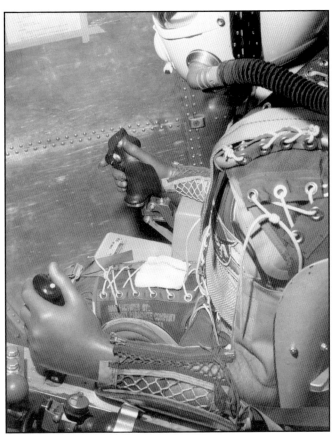

In the search for simplicity, this experimental suit didn't make the cut for obvious reasons.

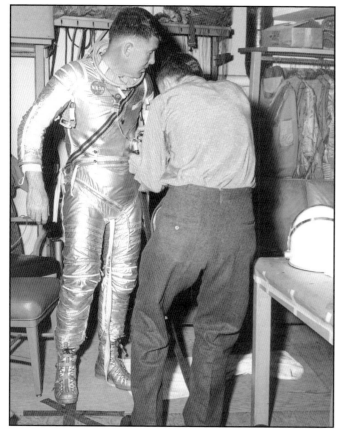

(Left) Mercury astronaut Walter M. Schirra, Jr., Lieutenant Commander, USN, shown receiving final adjustments to his Project Mercury full-pressure suit prior to entering the gondola of the human centrifuge.

The helmet developed for the space traveler was probably one of the most complicated helmets ever designed. It was made of a composite material which is much harder than motorcycle helmets. However, this helmet was much more than a head protector. It was, in fact, a mini-work station. The astronaut needed his hands for various performance tasks. To free his hands, some of his equipment was built into the helmet. There was a tiny adjustable microphone built into the mouthpiece through which he could communicate. It also included a thermostat to control temperature in the space suit. A thermister also sensed the astronaut's breathing rate and volume and temperature.

A headset was built into the helmet and enabled the astronaut to receive audio communications. They looked like ear muffs and served as ear protectors. They were adjustable which gave them a sort of flapping appearance. Inside the top of the helmet were sensors to measure brain waves. There also was a miniature switchboard with its leads monitoring biomedical measurements. The fact that the brain function of the astronaut could be recorded and transmitted across the miles was truly remarkable in the 1960s. Also attached to the helmet was a visor which enabled the astronaut to adjust light and to protect his eyes. Some early versions of the

(Left) Astronaut Gus Grissom, wearing a modified pressure suit, is pictured on his way to a centrifuge test ride to simulate a Gemini profile. His gloves – with special straps and wrist bands – were the result of much research. His ability to carry out piloting performance tasks when the suit was pressurized was the emphasis of these studies.

helmet protection even had two or three visors. They were eventually discarded because they were so heavy they often gave the wearer headaches. The visors were different colors depending on their use and visibility standards.

Helmets, as well as other clothing developed for the astronauts were made to protect them in case of a crash landing.

The specially-designed gloves might appear to be ordinary workmen's gloves. However, they had zippers and rings around the top and straps around the base of the fingers. They were made of flexible material so as not to restrict hand and finger movements. As anyone who has ever chaffed at the problem of trying to work with cumbersome gloves knows, gloves often make finger movements too clumsy. Workers on this planet often take off their gloves to solve the problem. Astronauts, on the other hand, can't remove their gloves even when they are required to press levers and to push buttons. Error-free pilot performance is essential to their survival, adding to the problem.

Another important part of astronaut's equipment was the biomedical sensors placed on the body to ensure that the life support system and the human pilot were functioning properly.

The first biomedical sensors measured heart function, blood pressure, pulmonary function, oxygen utilization rate and quantity, voice quality, body temperature, visual response, and sometimes brain function and muscle function.

This was done by attaching sensors to the body through a harness and undergarment which had a small electrical switchboard assembly attached to the pressure suit. The miniature switchboard contained 16 wire leads which measured critical aspects of the life support system and pilot performance requirements. Careful attention was paid to temperature readings from various parts of the body. This included both core and peripheral temperature. It was necessary to keep a constant temperature to maintain body comfort and safety. An astronaut could not be sent into an unknown environment and allowed to freeze his feet and bake his head—or vice versa.

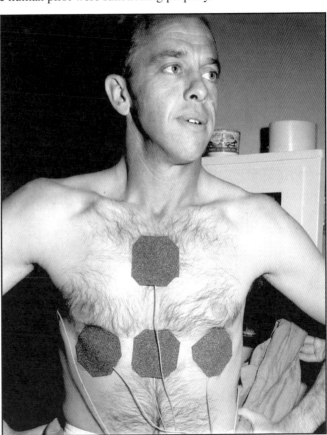

Ouch! Astronaut Alan Shepard dutifully wears a set of early biomedical monitoring devices glued to his chest for high-G runs.

Some of the data was monitored on displays in the laboratory as it occurred. Other data was recorded in vast recording systems and later intensively analyzed. There were recording sites in various parts of the laboratory building and in some of them data was transmitted to other research centers.

There was data describing medical, biomedical and physiological, psychological and performance, engineering, training, briefing and debriefing. Data had several forms: voice recordings on tape, engineering graphics on remotely-located computer recorders, and films of face and body and equipment during the tests; and ratings and comments by scientists and astronauts. Scientists love their data and seemingly no detail was overlooked.

They even reused the data in refresher training, for designing new training programs, and for progress reports.

Many of the biomedical sensors were miniaturized and redesigned for medical use and can be found making valuable contributions to the health and safety of patients today. Many of the findings from these studies also appear in science, bioastronautical, engineering, and aerospace medicine books and technical reports.

Another piece of protective equipment—which is beyond being placed in any kind of category—is the contour couch. This piece of equipment is a special kind of chair which was custom-fitted to each astronaut and played an important role in protecting the first astronauts during the gravitational stresses of launch and reentry. Repeated tests on the human centrifuge indicated that the astronaut could withstand much higher gravitational stress while strapped securely into the contour couch than he would have been able to endure without it.

Many of the earliest contour couches were made for research scientists who spent hours riding the human centrifuge and testing the couch in different ways. There was a room at the aviation medical laboratory where the couches were lined up like ghostly, alien creatures with hard shells. The process of custom-fitting each contour couch to the body shape of the researcher resembled some strange sculptor at work. The researcher or astronaut, wearing his flight gear, was dropped into a pit containing a sandy molding mixture. (The 6-foot drop would be enough to discourage most of us even before a couch was made from the mold.)

There were several types of couches made from different materials such as styrofoam, plastics and metals.

The couch was molded not only to fit the astronaut but also to fit within the Mercury capsule. The contour couches were remade for each type of spacecraft.

Not to be left out, the chimp astronauts also had their custom-made contour couches. (It is predictable that scientists being the researchers that they are, would then compare the two contour couches in an animal-human project.) Contour couches were used in the Mercury, Gemini and Apollo research projects and flights. However, they became too bulky for the sleek space shuttle flights and were greatly altered. Astronaut Steve Hawley, attending a space meeting with Dr. Chambers in 2004, said that the old contour couches looked a lot more comfortable than the newer version that was used in his Hubbell spacecraft. Contour couches had restraint systems for the head, the chest, the hips, the legs and the feet. The arms were free to make movements.

The elaborate restraint system was out of respect for the 17,000 miles-per-hour speed required for getting off the planet. Until the dawn of the space age, no human had ever been subjected to such speed. Many of the developments for astronaut restraint systems have been utilized for earthbound transportation systems. Restraint systems for airplanes, trains, cars, busses and boats have all been impacted by space research. The protective suits developed for astronauts have also helped to save the lives of firefighters, high wire electrical workers, tree trimmers, bomb squad and hazmat workers, and others who work on weather and rescue operations. The entire field of safety engineering has been greatly enhanced by the diligence and perseverance of space researchers. Medical science also has utilized many of the space research developments.

Scientists from at least two dozen disciplines had a hand in preparing humans to get off the planet and return safely. One little-known group was the environmental scientists whose earthly mission is to make and keep the environment safe and comfortable. Outer space is neither safe nor comfortable, so making space cabins livable was a tall order.

Even such ordinary activities as eating, drinking, bathing, sleeping and even breathing, presented complicated problems for space researchers. The fact that no human had ever ventured into that mysterious, hostile environment was especially challenging. If we made a list of the basic minimum requirements for health and comfort, outer space would have none.

Supplying "climate controlled" cabins was nearly impossible. Forget the thermostat on the wall keeping a room comfortable. A space traveler encounters wild fluctuations of temperature, the outside of his craft can sometimes be as hot as 6,000 degrees during reentry and as cold as 200 degrees below zero in some regions of outer space, depending upon the position of the planet with respect to the sun.

The astronauts' protective clothing as well as their vehicles had to be sturdy enough to see them through all of these harsh conditions. Their clothing also had to be comfortable, which posed difficulties because of the requirement that their pressure suits be sealed so tightly.

The atmospheric pressure factor dominated much of the research on habitability as it has in other projects. Providing breathing air as well as ventilation for the cabin was complicated. It was essential to transport an adequate supply of breathing air and a system which was leak-proof. Lost luggage is frustrating to any traveler. But the loss of breathing air by a space traveler would be catastrophic. Noise was also the subject of much research. Most of us pay little attention to the science of acoustics unless we find ourselves in a building where the noise factor has been neglected. The scientists had to protect the astronauts' hearing from the deafening roar of launch and reentry. During those phases of flight the space traveler is subjected to noise and vibration that is so intense it interferes with pilot performance. Then comes the challenge of providing comfortable acoustics during the many other phases of space flight.

In outer space the acoustical challenges change abruptly. Outer space has no noise so the astronaut has to adjust to a lack of noise outside and the sound of his equipment operating in the cabin. The cabin was outfitted at various times with heaters, air conditioners, computers, radios, fans, microwave ovens, microphones and instrument panels which sounded much louder than usual in the soundless environment. As unlikely as it may seem, the lack of noise in outer space presented problems rarely before encountered. There are so few places on Earth that are super quiet.

The human ear has no "off switch." It is always tuned in and turned on to sound. The question then arises how human performance, an essential ingredient to a successful flight, will be affected by this phenomenon. There was also the question of how the varying conditions of high-G and microgravity affected hearing. The ability to hear and understand messages sent by mission control was essential.

The scientists looked to their usual assistants—simulators—to study noise and the lack of it. One simulator was an anechoic chamber in which sound could be controlled. Another acoustical simulator was the "woofer" which could simulate the noise produced in launching a spacecraft and also simulate a sonic boom. There were noise simulators at the U.S. Naval Air Development Center and others at NASA Langley, Stanford University and North Carolina State University.

Scientists also studied the long-term effects of space conditions on the ear. It was concluded in centrifuge studies that pilots could hear well enough to press the abort button up to 14-G for a few seconds.

Other studies of the effects of space conditions on hearing were done by placing guinea pigs in the woofer and studying their hair cells in the inner ears and semi-circular canals. The studies of the effects of varying levels of noise and quiet on human performance generated a great deal of interest by the industrial world. In the early days of aviation, the "woofer" was used to study the effects of airport noise, such as the sonic boom, in the planning of airports and the design of airplanes.

In the 1950s when passenger airplane routes were being established, noise was a big factor to residents who lived near airports. Farmers blamed the noise for reducing the productivity of their milk cows or their laying hens. City folks pointed to cracked window panes and sleepless nights. That sounds strange in the 21st century where it is hard to find a quiet place and few people seem to expect it.

Another aspect of livability taken for granted here on Earth is lighting. We have become accustomed to the lighting of our choice at the touch of a switch. Lighting in a spacecraft, like everything else, was far more complicated. In a 100% oxygen partial pressure environment some of the electrical connections in the light fixtures had a tendency to explode. There were also great extremes of light and dark in the space environment depending on the position of the sun. The scientists tried all kinds of lighting for different spaces and function. They also developed eye protectors and visors for the astronauts in extremely bright light. Lighting for interior cabins and instruments evolved mostly from aerospace technology. The main problem with lighting was that the system had to be adapted to each different vehicle and mission requirement. Lighting is also subjective. The brightness level and angles at which to read the control panel differed widely among the astronauts. A camera was placed in the cockpit to give an objective reading of the panel lighting and to make a comparison between human vision and camera or robotic vision.

This was complicated by whether they were on the dark side or bright side of the Earth with respect to the sun. Other factors were the design of the helmet and the effectiveness of the visor. There was also a question whether the eye would continue to function in microgravity and in high-G Scientists knew the ability of the eye to focus would be different but the question was by how much. Some researchers feared that space myopia would hamper pilots from making spatial judgments outside the cabin window and also inhibit their ability to take pictures and to read the instrument panel.

Food preparation has a complex history of its own. In the earliest Mercury flights the astronauts started the day with a high protein steak breakfast and carried no food with them into space. The purpose of this plan was to allow the scientists to study the effects of space flight on digestion and the utilization of food. The opportunity to study how the digestive system functions in the microgravity environment had never been available before. The astronauts' diet was carefully controlled while scientists carefully analyzed the body's utilization of many nutrients. Biochemists also analyzed the red and white blood cells as well as other blood components.

Urine and feces were also analyzed after a disagreement between some of the astronauts who wanted to dispose of body wastes in space and the biochemists who insisted they bring it back with them for study. In the midst of the digestive studies Astronaut John Young received wide publicity for smuggling an unauthorized super-sized sandwich aboard one of the Gemini flights.

After a good meal many of us like to sit in a comfortable chair. Comfortable seating was especially difficult to attain in a spacecraft because of the cramped space and long hours of confinement. Carrying an adequate supply of pure, palatable drinking water was not a problem for short flights. However, it was a significant problem for longer flights. In the early days researchers tried to utilize foods that contained a lot of water to limit the amount needed to be carried separately.

Another big question was how to drink the water both in high-G environments and in weightlessness. In both cases the problem was to keep the water in the container and also accessible. They couldn't tolerate having water molecules drifting around in the cabin. Engineers, nutritionists, biochemists and physicians all contributed their expertise to solving the problem. The end result was a variety of bottles and squeeze-tube containers. These had special valves in the top which enabled the astronaut to drink without dispersing the water into the cabin. At one point in research the scientists worked on the idea of reusing drinking water and thereby conserving it. Several models of various contraptions were built which would salvage urine and through a series of filters use it for bathing and ultimately reuse it for drinking.

Some researchers even had the idea that this type of water conservation equipment would be a boon to hotels and cruise ships. Even though urine could be purified enough to be acceptable by some, it was never popular. Astronauts, who ordinarily made many sacrifices of their creature comfort, drew the line at this one.

Another far-out idea put forth by some creative thinkers was to develop edible clothing and equipment. The idea was that this would conserve precious weight and space as well as provide emergency food. This sounds less bizarre when one considers the possibilities. The early astronauts could not be guaranteed a landing in a precise spot. In fact, some of them landed far from the planned landing site— once even in the wrong ocean. So having food to tide them over in case of a delayed recovery had merit. The idea of eating your shirt was never tested but may still be considered for the long trip to Mars.

The question asked most often by school children—and others—is about space toilets. This was a subject of great concern, research and innovation. The first waste disposal systems were bags attached inside the space suit and used for short flights. Then specially modified toilet containers were developed. In the weightless environment toilets were fastened inside the space capsule and a system of pumps, valves and fans forced the waste into sealed containers.

In the life support lab at NASA Langley, more advanced waste disposal systems were developed. The first was a soft-sided compartment which later was enlarged to encompass the whole body. Different versions of these were built and tested by several industries as well as government labs. The consensus was that when it comes to bathrooms, there's no substitute for regular plumbing and there's no place like home. The same can be said of showers in space. Forget bathtubs. Forget hot water and bubble baths. The earliest astronauts had to be content with sponge baths. Showers were developed for the shuttle and Sky Lab, the U.S. space station of the 1970s. There were several designs for showers. They all involved having the astronaut climb into a large bag or compartment. It was sealed to prevent moisture from drifting around in weightlessness. However, it also had to provide ventilation.

Heating water was limited because of the requirements for conserving fuel, so showers typically were taken with water at room temperature.

Even the simple act of brushing one's teeth took on new challenges in outer space. After some research and innovation, the method developed was to brush one's teeth with a closed mouth and with edible toothpaste which could then be swallowed. When it comes to bathrooms, many improvements have been made but space facilities still are far from the conveniences to which most of us are accustomed. Even the hardy souls who enjoy wilderness camping have more luxury than space travelers. One of the most unusual aspects of early astronaut training was the effort invested in keeping the pilot alert throughout the flight. Most of us enjoy snuggling down in our comfortable beds for uninterrupted hours of sleep. This luxury was not an option for the early astronauts.

It was imperative that the astronaut not allow himself to doze off into deep sleep. He had to stay alert enough to hear and respond to mission control communications and to monitor the complex array of instruments, controls and displays. The task was daunting and crucial. Every flight required the pilot's numerous split-second responses performed in exact sequence, all of this while coping with acceleration stress, reduced atmospheric pressure or the exotic environment of weightlessness.

The pilot also had to be vigilant for any malfunction of any of the complex equipment and so to be prepared for a split second emergency abort, if necessary. In other words, a spacecraft is not the place for a drowsy pilot.

In March, 2005, pilot-adventurer Steve Fossett set three world's records for flying solo non-stop around the world in 67 hours. He reported that he had slept in two-minute increments which evidently sustained him. During alertness training the astronaut learned to focus on the task at hand and to stay alert no matter how he felt. A number of psychologists worked at developing alertness exercises with which the astronaut could test himself during the flight. One prominent test was a running matching memory test similar to board games requiring pairing similar shapes, numbers or symbols.

Another was a specially-designed instrument called the Langley Complex Coordinator which measured reaction time along with skills required to coordinate eye, hand and foot movements for complex tasks. A miniature version of this was placed on some of the early flights.

Each of the early flights was preceded by mission simulations on the human centrifuge in which they underwent alertness training. The first space flights were short and the whole idea of actually exploring such a strange, new place was so exciting that they probably couldn't have slept anyway.

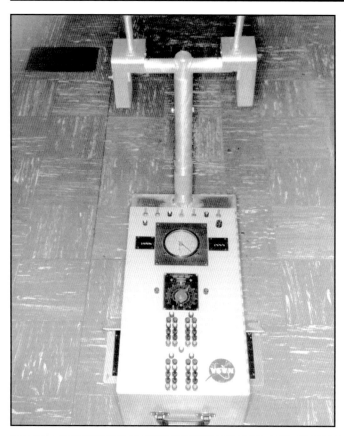

A serious problem for the early flights of long duration was how to keep the astronaut alert. Most of this alertness training was done on a spacecraft instrument panel. This Langley Complex Coordinator was designed to help the astronaut check his alertness state. It was tested during the 90-day space station simulation. A portable version of the device was designed to be carried aboard the spacecraft.

(Below) This researcher illustrates how the device is operated. It may eventually be adapted for medical use such as testing and evaluating stroke patients.

The Gemini flights with their two-man crew could alternate sleeping but they were confined to such a small space that truly restful sleep was impossible. The Apollo flights with their three-man crew were also cramped for space. The first real "beds" were a variety of hammock-like designs for the shuttle and Sky Lab. With large crews they could rotate duty and sleeping. Sleep patterns varied with each astronaut as well as with flight requirements for performing their missions. The purpose of the flights was to explore this strange new place and to report what they saw in the ocean, on land and in space.

Vast expanses of sky or ocean often became monotonous and tended to make the astronaut sleepy. However, this was counterbalanced by the excitement of riding a rocket into the vast unknown regions of space, which was so exciting to some that it left them sleepless. The snugly-fitting pressure suits, the neck ring, gloves, boots, and biomedical monitoring instrumentation as well as confinement of the contour couch in the small cabin made sleep difficult and uncomfortable.

The contour couch, custom made for each astronaut, was adjustable like a recliner and could be turned into an upright chair. Once in outer space the spacecraft could be flown at any angle which was comfortable. The problem with that was that sometimes the astronauts disagreed about which angle was best because comfort is a very subjective matter.

For example, the length of the flight was a major consideration. A contour couch that was comfortable for one orbit was not necessarily comfortable for 24 hours. Boots that were comfortable for a short flight began to pinch after three days. Even though the space travelers might not have believed it, the space researchers spent much time and effort on the subject of comfort. The form-fitted Mercury contour couches, for example, were tested and evaluated by flight surgeons and psychologists and rated for subjective comfort and ride quality.

Testing of this contour couch is being monitored here by Dr. Chambers. It was designed for comfort during high-G runs.

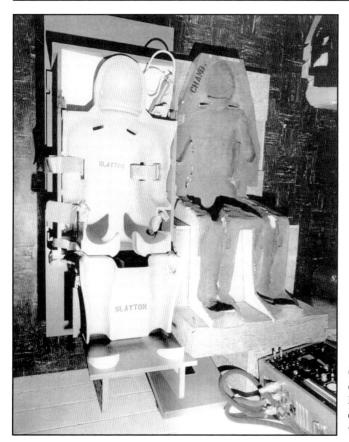

Despite all of the research efforts and innovative equipment, travel into outer space still is not a luxury cruise! In fact, anyone who aspires to space travel has to be willing to give up practically all of our modern-day creature comforts for the adventure of it.

(Left) These contour couches could be called the scientists' version of the recliner. Researchers devoted a lot of time to cabin comfort, although the astronauts probably would have trouble believing this.

Most people probably thought that the Mercury flights off the planet would make subsequent flights easier. Like most other things in space research, the situation grew more complicated instead.

The human factors research had to be repeated for each type of vehicle, with each increase in crew size, with each change of rockets and with each mission.

In recent times many people seem to have a disregard for detail, preferring instead to sum up a discussion with the word "whatever." That word is not part of a scientist's vocabulary. They give new meaning to the phrase, "a stickler for detail." Not only would scientists insist on researching every unknown aspect of a project, they would repeat the study for confirmation. Consequently, they would never send an astronaut into space without having thoroughly studied every factor in every move. Merely hoping for the best was not even an option.

Another little-known fact about the early flights is that scientists were researching variables for the moon walk even in the Mercury and Gemini series. Space technology was in the early stages of development. Technology for the Mercury flights had to be developed and applied before moving on to more complicated flights.

Technology for boosting the spacecraft off the pad as well as developing technology for operating the launch site had to be developed. This included the countdown procedures, for lift off and guidance into the Earth's atmosphere and space. The guidance systems received a great deal of attention from systems engineers, astronauts, flight crews and world wide monitors. As the technology for guidance systems was developed, both the ground crews and the flight crews had to be trained. Inertial guidance systems were developed which held the spacecraft on the precalculated flight path. This was augmented by good old-fashioned optical navigation, gyroscopes and accelerometers.

Both the ground crews and the flight crews had to become accustomed to travel speeds never experienced by humans before: orbiting the Earth in 90 minutes, being over the Atlantic Ocean one minute and passing over Australia soon after.

The inertial guidance system, improved from older technology used for ships and aircraft, was a marvel in itself. It provided navigation, guidance and control of the spacecraft, measurements of vehicle performance and environment, transmissions of data to ground control stations and vehicle monitoring sites. There were even sensors which detected bad weather, foreign objects or hazardous conditions. Astronauts were trained to manage the inertial guidance system and practiced sequencing and decision strategies. During both the acceleration stresses of launch and the clueless conditions of weightlessness the astronaut had to learn to keep the spacecraft in a proper position in relation to the flight path.

Training technology was also developed for each space mission of Mercury, Gemini and Apollo. Many space scientists, including Dr. Chambers, contributed to the writing of training manuals which were similar to an owner's manual that comes with a new car. They also developed training devices.

Many astronauts spent at least 300 hours in training and preparation for some flights.

The first studies—in the simulators—were designed to ascertain if human beings could cope with physical forces such as acceleration stresses, atmospheric pressure changes and weightlessness with protective clothing and equipment and training. They not only had to survive but also to perform piloting tasks. This data was also used in the design of the spacecraft. After Mercury, Gemini and Apollo capsules were produced, there were more rounds of research and training to acquaint the astronauts with their vehicle and to train them to operate it. The training also included the support groups on the ground.

The skill training programs actually came from earlier studies such as those by Dr. Chambers when he was assigned to the flight component skills laboratory at Lackland Air Force base when he was on active duty in 1955.

Of course, some of the basics of training were based on older studies and data.

The early Mercury flights were used to meld all of the elements of man-machine and pilots and ground support crews into a skilled, efficient system dedicated to performing the mission safely.

The era of Mercury flight preparation was also one in which great improvements were made in communications and computer technology. Younger generations with their palm-sized cell phones and watch-sized computers probably have difficulty even imagining the state of these devices 45 years ago.

At that time the computer which operated the centrifuge occupied a large building. It was considered to be very complex and high tech, even though it was very temperamental and unreliable. It had enough wires extending from its mysterious inner workings that it looked as though they could reach the moon without a spacecraft!

Communications in those days were conducted mostly by wired telephones, telegraph, and short-wave radios. Astronauts could not be sent into outer space without communications and in the 1950s and 60s it looked as if the communications scientists had their work cut out for them. This was also true of the specialists working on the biomedical monitoring devices. The first biomedical monitoring was very clumsy and uncomfortable.

The first civilian astronaut, a young aeronautical engineer and test pilot named Neil Armstrong, joined the astronaut corps (Randy had worked with Neil during the human factors work on the X-15 aircraft and also when Neil worked as an aeronautical engineer on the Mercury program. Everyone was impressed with Neil's knowledge and abilities.) Despite Neil's exceptional qualifications there was grumbling among the others, each of whom had hopes of being the first man to walk on the moon. Many observers felt that the competition among the various military services to have their representative be the first moon visitor was the reason a civilian was appointed. Under this arrangement, none had to choose among the Navy, the Air Force, or the Marines. Thus no bureaucrat had to endure the wrath of admirals or generals in high places.

Gemini—with its 2-man crew—was used to further study the cooperation and coordination that would be required for a moon landing. Many of the experiments and training exercises done singly were now repeated by pairs of astronauts. After much research by the behavioral scientists, the first EVA (extravehicular activity) was conducted on Gemini 4. This was essential preparation for a moon landing and there were many aspects to be studied. Being launched into outer space at speeds heretofore never experienced by humans and with perils too numerous to list, was one thing. But climbing out of your spacecraft and finding yourself all alone in the vast reaches of outer space was something entirely different.

Researchers began by studying Earth separation phenomena. This is a feeling of uneasiness and separation often experienced by submariners or Arctic explorers who live in isolation for long periods of time. To a lesser extent Earth separation affects people who are moving to a different place or those whose familiar environments are drastically changed by floods, hurricanes or other disasters. However, none of these measure up to an EVA in which the astronaut really is separated from his earthly home and attached to his spacecraft—his lifeline—by only a tether.

Each person reacts to Earth separation differently and some become so nervous that it may impair their performance or make them physically ill.

At the other end of the scale are space explorers who experience euphoria similar to "rapture of the deep." This phenomenon, observed for years in some deep sea divers, excites the explorer so much that he doesn't seem to care if he never gets back home. Someone experiencing this feeling wants to continue beyond the limits of safety set up for the mission. For example, one of the astronauts began arguing with ground control when it was time to end his EVA. Astronaut Ed White made the first American EVA aboard Gemini 4 in 1965. It lasted a full 20 minutes with the astronaut suspended outside the spacecraft by a gold-coated "umbiblical cord." There was also great concern by the astronauts, the researchers, monitors, flight controllers—everyone involved—that all of the equipment work without fail. Life support equipment, video systems, flight control equipment and communications systems were of special concern. A unique type of flight controller was introduced for the EVA missions. It was a hand-held instrument, about the size and shape of a large flashlight, with which the astronaut could direct his movements while suspended in space.

What might be looked upon as the most daring EVAs were the later ones in which the astronaut unfastened his tether and used his hand-held flight controller to keep close to the orbiting spacecraft. (Sometimes they held a flight controller in each hand.) As risky as it sounds, this type of EVA was necessary to prepare for the moon landing.

There were a total of 10 manned Gemini flights ending with Gemini 12 in November, 1966. From the beginning the shadow hovering over space research was the possibility that some experiment would go terribly wrong with fatal results. The specter of fatalities among the brave young astronauts haunted the researchers for years.

Then in 1967 it happened. Gus Grissom, Ed White and Roger Chaffee perished in a fire at the Cape during a pre-flight rehearsal for Apollo 1. The irony of it all was that it happened on the ground—not in orbit as had been feared.

For several weeks the news featured the sorrowful pictures of widows and children with flag-draped coffins and buglers playing taps at solemn ceremonies. We wept with them all.

We also wept for the death of the dream that the space program could be carried out with utter safety and without casualties. Many of the space research efforts were frozen for at least the next two years. Numerous investigations were conducted by a variety of governmental and scientific committees and panels. Changes were made in Apollo plans. Several flights were canceled outright and others were changed to include more investigation of the moon's surface. There was a renewed determination by the scientists to investigate every possible hazard, known or unknown, that might be encountered on the moon's surface. Some of the numbered Apollo flights were unmanned studies of various issues. Apollo flights 8, 9, and 10 were devoted to further data collection and analysis.

Then on July 20, 1969, Neil Armstrong and then Buzz Aldrin dared to walk on the surface of the moon while Michael Collins orbited the moon. When they placed an American flag there the world watched in awe.

The crew of Apollo 11 – Neil Armstrong, Buzz Aldrin, and Michael Collins – established their place in space history on July 20, 1969, when Armstrong and Aldrin landed on the Moon, while Collins served as the Command Module Pilot.

The scientific community—in a mighty cooperative effort—had fulfilled President Kennedy's mandate that Americans walk on the moon within the decade. This feat will stand as possibly the greatest scientific achievement of the 20th century. After Apollo 11 there were several more moon flights, each of which was remarkable and contributed to scientific knowledge about the universe. Then came Apollo 13 which has come to symbolize for many the entire moon exploration program. Its harrowing and heroic brush with disaster eclipsed all of the successful moon flights before and after.

It has never really been explained that the simulations done on an emergency basis to figure out how to bring Apollo 13 home safely had all been devised by scientists during earlier research and training studies. Nor has it been mentioned that a great part of astronaut training centered around coping with emergencies such as equipment failure. A big factor in astronaut selection focused on the pilot's ability to keep a cool head and to handle emergencies with calm and innovation.

After Apollo 13, it was almost as if the public began to expect more drama from the Apollo series. A successful flight soon became routine and boring to the public. In all, there were six successful Apollo flights which involved 12 astronauts who walked on the moon. Gene Cernan in Apollo 17 was the last man to have this experience.

Astronaut Eugene Cernan on Apollo 17 was the last man to walk on the Moon. He is pictured as he saluted the flag at the Apollo 17 landing site.

Movies such as "The Right Stuff," "Apollo 13," and various science fiction series all have come to represent the space program in the minds of the public. If a survey were taken it would no doubt reveal that many people think that the actor Tom Hanks was one of the real astronauts. Space scientists were very disappointed when the moon flights were ended. They wanted to continue to reap the treasure trove of new information discovered with each flight.

Apollo 13 will be forever representative of the Apollo flights. It is indeed ironic that the flight that nearly ended in tragedy has overshadowed the tremendous success and the amazing achievements of the other Apollo flights. Part of the reason for this is that scientists, by their nature, are serious, introverted scholars and will rarely become entertainers. As I discovered some 50 years ago, the life of any party will almost never be a scientist. Even if they tried to tell a joke they probably would forget the punch line. In this world of glitz, glitter and sound bites, scientists will seldom be found in the limelight. Center stage is not their medium and drama is not their style.

I once heard a pilot describing his trip around the world with a plane load of celebrities. His account was so dull the audience nearly dozed off. I could find more drama in a trip to the post office! I think this is the reason that the real Apollo accomplishments pale compared to the drama of the movies. However, scientists probably like it that way. They're far more comfortable in their laboratories anyway.

Ninth Manned Apollo Crew – The members of the Apollo 15 prime crew are (left to right) James B. Irwin, lunar module pilot; David R. Scott, commander; and Alfred M. Worden, Jr., command module pilot. Apollo 15 was the fourth lunar landing mission and the first to use the Lunar Roving Vehicle (shown with crew) to traverse the lunar surface.

When President John F. Kennedy proudly proclaimed that the United States would put a man on the moon in a decade, it is not clear if he—or anyone else—realized the enormity and complications of such an endeavor. Most scientists are highly specialized and the space program required a wide variety of scientific disciplines. Another requirement—equally important—was that the scientists be imaginative and humble enough to admit when they lacked knowledge. They also had to be tough enough to withstand the barbs of the naysayers, who were numerous.

The manned part of the project—which Randy was helping to mastermind—was especially sensitive. In those days, rockets blew up with unfortunate frequency amid sighs of dismay and budgetary shortfalls. But putting human beings at risk was even more challenging. Vice President Lyndon Johnson cautioned the scientists that a loss of human life or even severe injury would be unacceptable.

The human being is the weakest part of the man-machine system. However, he is also the best problem solver. But expecting 100% reliability of man or machine would be an almost impossible achievement. The first simulator studies were conducted to find out if fragile humans could survive and function in such a hostile, mysterious environment. The human centrifuge studies tested the ability of astronauts to survive the gravitational stresses inherent in lift off, reduced gravity and reentry. Following that were the water tank studies and parabolic flight maneuvers which simulated weightlessness. If he was to return safely the human explorer had to be able to survive and work in ways heretofore unknown. The third basic simulator was the atmospheric pressure chamber which in many ways was the most essential factor to life scientists. As unlikely as this sounds to us non-scientists, what this amounts to is devising ways to make atmospheric pressure—the finely-balanced gasses which are essential to life—into a portable envelope in which the explorers can be encased while they leave the planet.

After all of this was worked out, there were even more challenges if astronauts were going to land on the moon. The engineering groups were hard at work designing the equipment but the human factors researchers were faced with even knottier problems.

Three vehicles would be required for moon landing and exploration: the Apollo spacecraft which carried the crew and the cargo to the moon, the LEM (Lunar Excursion Module later simply called LM) the smaller craft which would carry two astronauts—and later the Lunar Rover—to the moon's surface, and then the jeep-like Rover in which the astronauts could explore the moon itself. (The Lunar Rover was not deployed until later Apollo flights.)

The moon project also required the development of more powerful rockets than had been used in earlier flights. This involved rocketry engineering and also procedural training for both the astronaut and the ground controllers. The Redstone, Atlas and Titan rockets had been used to launch the early flights of Mercury and Gemini. However, the moon flights were to be launched by the Saturn rocket, a much larger and more powerful rocket. Human factors scientists and engineers were also called upon to pioneer procedural training to ready the astronaut for controlling the mighty rocket.

If the moon landing were to be successfully carried out, the crew of three and their equipment all had to be operated perfectly in precise split-second timing for landings, rendezvous and docking. And once again the scientists turned to simulation as the best way to prepare for such an endeavor.

Procedures training—which could be compared with "Driver's Ed"—was provided by installing a procedures trainer device in the pressurized centrifuge and simulating the Saturn lift-off and separation. Both astronauts and ground flight controllers spent many hours practicing intricate procedures involved in controlling the gigantic rocket.

Atmospheric pressure—which decreases as altitude increases—affects hardware as well as humans and this also had to be reckoned with. The procedures for managing cabin pressure and suit pressure were rehearsed inside the atmospheric pressure chambers until all of the knotty questions were answered.

The human centrifuge was then reprogrammed to simulate the cockpit, controls and displays of the Lunar Excusion Module, both when it was inside the Apollo spacecraft and then when it would be used to perform lunar landing maneuvers. The government scientists were working closely with scientists and engineers from the industries which were manufacturing the craft and instrumentation.

Key maneuvers required for a successful flight were the ejection maneuver and then the lunar landing. This involved a series of flight task procedures which had to be carried out with great precision and in proper sequence.

A wristwatch designed for this project graphically depicts the complications involved. The watch has four faces. It not only shows the time in various parts of the world, but provides a running log of the elapsed time from lift off, the moon visit as well as readings of the fuel levels. Unlike earthly travelers who leave home and then return days or even weeks later to the same landscape, the astronauts must adjust to the ever-changing heavens. Thus the watch with four faces helped to coordinate the entire project for the astronauts, the equipment, the ground communications and the ever-changing position of the moon, the stars and the spacecraft. Back-up support also was supplied by equipment in the capsule and ground control.

To the untrained eye the LEM looks somewhat like a large bug or a robot like the one later sent to explore Mars. The first maneuver to be practiced in the LEM was the ejection from the Apollo craft as it orbited the moon. This was rehearsed by placing the astronaut inside the centrifuge with parts of the LEM. The LEM was too big to fit inside the centrifuge, so only parts of it—such as controls and displays—were simulated at one time. Mastering this maneuver involved both the ejection controls and the descent engines. This was more daring than any earthly stunt even daredevils could imagine.

The LEM—with its two passengers—was to be disconnected from the Apollo spacecraft in lunar orbit for the first time ever. (The third Apollo astronaut continued to circle the moon in the spacecraft in preparation for picking them up again for the trip back to Earth.)

The LEM was to be ejected about 100 miles above the moon's surface while the Apollo craft was traveling at around 15,000 miles an hour. It had just enough fuel of its own to descend to the moon's surface below. With the slightest error, the LEM could miss the moon landing site and be lost in three-dimensional space. The procedures for landing on the moon were practiced extensively in the centrifuge. The two astronauts in the LEM practiced both together and one at a time. The crew for the moon flights had not yet been selected so numerous astronauts from the pool took the training. They performed simulated missions and the crews were selected from these tests. The simulation studies were also used to determine which crew member should perform specific roles and tasks.

The surface of the moon was quite rocky with patches of soft, unstable composition. It was essential that the pilot find a solid, level landing site. If the LEM were damaged in landing, there would be no way to return to the orbiting Apollo craft. The atmospheric pressure and temperature factors also became noteworthy again. The lunar orbit—the pathway which the moon travels—has its own specific microgravity field with conditions very different from Earth.

This meant that many of the pressurized centrifuge simulations which had been conducted for other environments had to be repeated to study the microgravity of lunar orbit. The scientists could not take a chance that the first astronaut to land on the moon be felled by some unanticipated environmental menace.

They also practiced return maneuvers. This involved lifting off the moon's surface by firing the ascent engine on the LEM to meet the orbiting Apollo craft and then, with great precision, docking with it for the ride home. There was still another simulator to be mastered in the preparations for the flights to the moon. After practicing thoroughly in the Apollo spacecraft and the LEM, the astronauts were now faced with the prospect of stepping out of the LEM onto the strange, mysterious environment of the moon's microgravity. To practice walking in an environment which has 1/6 of the Earth's gravity, the engineers had designed a simulator called the Lunar Lander, one of which was located at NASA Langley in Virginia and later another was installed at NASA Houston.

The astronaut was suspended by large cables which enabled him to walk sideways on the wall with his feet pressed against the wall at the required 1/6 of the Earth's gravitational force.

A picture of this experiment was a sight that was unlike anything seen on Earth or any acrobatic stunt. With a little practice, the "moon walker" could skip and hop against the wall as they later did on the moon. They also practiced carrying objects to prepare for gathering rocks during the moon walk.

A new challenge was the Lunar Rover which was not used until Apollo 15, although much detailed research had been done with it for many years. Training the astronauts to manage and work with the Lunar Rover was like getting acquainted with a new car while learning to function in microgravity. The Lunar Rover was transported to the moon aboard the LEM, folded up like a child's stroller to be as compact as possible. It was to be used to explore the mountainous part of the moon where the astronauts were to collect rocks and other specimens for geological studies.

Learning to drive the Lunar Rover at $1/6$ of Earth's gravity required special training. This training was performed at simulators at Langley NASA, Houston and the New Mexico desert and at several company facilities. The main problems were the power supply and staying upright on the lunar terrain. (An earthly challenge similar to driving on the moon would be to drive a car on sand while carrying rocks, soil samples and equipment.)

In exploring the Moon's surface, Astronaut Harrison Schmitt collected many rock samples. These included samples from a huge bolder in the Moon's Littrow Valley.

The Lunar Rover enabled the moon visitors to explore a much wider area of the moon and to fulfill the mission to explore the moon's surface. This assignment kept them very busy photographing, recording and collecting samples of their exotic environment. The explorers also had to be very careful not to get hung up in the sand or to damage the vehicle on the rocks. The astronauts had to launch the LEM for the return flights from a specific site. If they had crashed the rover at some distant site, it might have been catastrophic.

The Lunar Rover was a marvel of clever engineering. The requirements were staggering: it had to be portable and light—but also sturdy. This amazing vehicle has rarely been fully described, especially for us non-technical types.

The Lunar Rover looks like a combination of a golf cart, a jeep, and a farm tractor, although it is much more complex. It also has unique features unlike any earthly vehicle. Prominent among these is an umbrella-like attachment which was both a solar-energy collector and a communications device. It also had specially designed steel-belted tires. (There was no room in this project for a flat tire! Also pressurized tires would have simply exploded in the vacuum of space.)

The Apollo 15 astronauts used the Lunar Rover as planned and with flawless performance. They explored the mountains of the moon and gathered rocks for study by the geologists who were eagerly awaiting on

Earth. The rover was "parked" on the moon and was also used by the astronauts of Apollos 16 and 17. The longest recorded trip for the Lunar Rover was 66 hours and 54 minutes of moon exploration.

The LEM crew had to be prepared to work in the untried, unexplored, lunar environment. To practice these tasks, it was back to the water tank simulators. This time the subjects were submerged in water and given numerous tests to give them practice in all manner of skills that might be required of them on the moon.

From the beginning of the space program, Randy brought home his work and also some of his colleagues. Many of them were far from home with little to do in their off hours. Our house became a gathering place for the visiting scientists, aerospace engineers, flight surgeons, test pilots and astronauts, who liked to talk about various aspects of the space projects. One of our visitors was a young aerospace engineer-test pilot named Neil Armstrong. Randy had met him several years earlier when both of them were assigned to work on the X-15 aircraft. Neil became one of our frequent visitors and he and Randy often talked about space research into late night hours.

Sitting in rocking chairs with cups of coffee or tea, the two exchanged views of the space program and its possibilities. One night after I had gone to bed Neil evidently broke one of my cups. As the mother of two active boys, I had accepted a certain casualty rate of household breakables and I soon dismissed the incident.

I was surprised several days later when a deliveryman from a local department store presented me with a box containing two cups and saucers. Neil had thoughtfully ordered a double replacement in my pattern.

Of course in those days we had no idea of the fame that awaited him. I have always regretted not putting a special mark on the replacement cup.

On one occasion we had a particularly stressful week at the acceleration lab—equipment failures; on again, off again projects. Disagreements among the various specialists. By Friday night a group of six or eight scientists, engineers, test pilots and two or three astronauts gathered at our house for dinner.

After dinner—as if propelled by a need to let off steam—we put together an impromptu band. Randy and Neil Armstrong played trumpets. Others picked up toy drums, a toy xylophone, several kazoos and pots and pans. I "accompanied" this group at the piano. (I should hasten to add that I practiced the piano because it was such a challenge—not because I had any latent musical talent. Actually, few people have practiced so hard and played so badly.) We played our old favorites, such as "She'll be Comin' Round the Mountain," "On Top of Old Smokey" and "I've Been Working on the Railroad." We were loud—and dreadful. Most of our renditions were hardly recognizable. I laughed so hard I almost fell off the piano bench. Our spirits were considerably improved after our session of silliness. We laughed again when we realized that we had probably reinforced our neighbors' impressions that we had a strange lifestyle.

The Lunar Rover – which looks like a combination of jeep, golf cart and farm tractor – was a remarkable engineering achievement.

One time Randy went on a trip, delivered to a conference by a test pilot. He called the evening they were to return home to report that the plane had engine problems and they would be delayed. (I learned later that a piece of the engine was missing!) When he finally arrived home the next afternoon I asked, "Did you have trouble getting the plane fixed?"

"Oh, we never got it fixed," Randy replied, "we just decided it would be better to fly in the day time."

I didn't understand that kind of thinking until about 40 years later. I met Alan Shepard's daughter at a gathering at the Kansas Cosmosphere. She told the story of how her father, originally a test pilot, defined the job: "a test pilot takes off in a plane that needs repair so he can find out what's wrong with it."

In the 1960s most homes did not have central air conditioning. During one of our summer dinner parties a fan tumbled out of the window and broke. We had four engineers in the group—mechanical, aeronautical, human factors and one who specialized in projects. These were highly-trained engineers who were helping astronauts and spacecraft get off the planet. My first thought was that surely fixing a fan would be a snap for these men.

Not so. One by one the engineers failed, announcing, "I'm not that kind of an engineer!"

The fan was finally fixed by a flight surgeon and a research psychologist.

For most of us, our lives are measured by milestones such as graduations, weddings, birthdays and anniversaries. Over the years the space flights became personal milestones at our house.

When Gus Grissom, Ed White and Roger Chaffee were killed in the Apollo 1 fire in 1967, we all felt the loss as if they had been family members. The disaster was especially devastating because Gus Grissom and the others had been such outstanding astronauts and this was the crew that had been selected to go to the moon. Gus and Randy also had spent their boyhood in neighboring parts of southern Indiana. Some thought the space program would never recover. It did take several years.

July 20, 1969, became a memorable date. When Neil Armstrong and Buzz Aldrin landed on the moon, the event became worldwide headline news and coverage.

But none of it was as personal as it was for the families of the astronauts, as well as the scientists and engineers who had worked on the project so long and diligently. Randall Chambers reacted to the moon landing like the coach of a team playing in a championship game. Would his charges remember their training? Would the equipment operate flawlessly? If not, would the carefully-practiced emergency measures be remembered and carried out?

Randy alternately paced the floor and tried to record the news reporters' commentary. Sometimes he seemed to be holding his breath until a specific maneuver was accomplished. It was hard to watch—but impossible not to watch. This, then, was the ultimate test, the culmination of the years of planning, research, development and training. Randy—and many of the other researchers—were probably searching their memories and reviewing the various programs to reassure themselves that nothing had been overlooked.

Some thirty-five years later, Randall talked with Astronaut Buzz Aldrin at a Space Development conference in Denver. Not only did the astronaut remember the human factors scientist, he chided him for not giving both astronauts a clear window view and instrument display in the LEM!

Two puzzling factors stand out when one discusses the moon walks. One is how soon they became everyday events. After Apollo 11 and the enormous publicity surrounding Neil Armstrong walking on the moon, the only project which commanded much worldwide attention was Apollo 13, which narrowly escaped disaster.

Apollo 14, 15, 16, and 17 seemed to be almost unremarkable. Another astonishing fact to earthlings was the interest shown by test pilots and astronaut candidates. Even with all of the perils, there were several thousand applicants for the Apollo moon flights and many were disappointed when they were not selected. It had truly been a remarkable decade between the time the project began and the moon walks.

In fact, it would be difficult to find another decade in history in which so much knowledge was discovered in such a short time.

Mrs. Chambers stitched this sampler as a gift for Dr. Chambers to commemorate Neil Armstrong's Moon landing. After that she gave up sewing in general and the cross stitch in particular!

We moved to Newport News, Virginia, in 1968, when Randall was assigned to NASA Langley as Chief Life Scientist and head of human factors engineering, a GS-15. This NASA division had as its mission the designing of new types of spacecraft and preparing astronauts to operate the new vehicles. After 10 years at the Navy's aerospace medical research laboratory in Pennsylvania, I was happy to leave the human centrifuge which had dominated our lives for so long.

However, it soon became apparent that this was turning out to be a case of "out of the frying pan into the fire!" NASA Langley had even more simulators. We had left the human centrifuge and moved to a brand new lunar lander simulator, a variety of atmospheric pressure chambers, assorted water tanks and wind tunnels, and noise simulators including the woofer. There also was an air bearing simulator and flight simulators used for microgravity studies.

(Randy brought home a table-top model of the air bearing simulator and it made me dizzy just to watch it spin like a drunken top.)

Also included in the research collection were various airplanes which were used in training and research. Training devices and other training methods being tested for use in space could be tried out first in these airplanes. Clearly, this was "research central!" It gave me hope that our new neighbors might also be better acquainted with space scientists and their unconventional ways. That hope was short-lived.

I soon discovered that NASA scientists had developed a reputation among the townspeople which preceded Randall's arrival. Shortly after we moved to Newport News Randy and I went to a flooring store and chose some new carpeting. After we made our selection, the salesman asked Randy where he worked.

When Randy replied, "NASA" the salesman startled us both by blurting out, "It can't be! I've had numerous engineers from NASA as customers and they always whip out a slide rule and start refiguring my measurements."

Then he added, "You're too normal to work for NASA."

What he didn't understand is that human factors engineers were busily measuring human capabilities with their slide rules. Slide rules are probably unfamiliar items to today's budding engineers and scientists. They probably would also be disappointed to learn that NASA scientists had such old-fashioned items in the 1960s. NASA did have both analog and digital computers in those days but they were large, many of them occupying whole buildings.

In the 1960s the slide rule was a favorite high school graduation present bought with great hopes by relatives of the new grad. Fast high-tech digital computers have relegated the slide rule into the proud history of aeronautics and space.

Our sons were 11 and 14 and took to their new shoreline environment like a couple of ducks. With the usual resiliency of youth, they were soon settled in their new schools and enjoying sailing and other water sports.

Randy and I had had a lot of practice at relocating by this time and we were soon putting down our roots and participating in church and community life.

For Randall, the shuttle represented a brand new research project almost like starting astronaut training all over again. The mission of space research had also been expanded. The Mercury flights had tested man's abilities to travel into space and to return safely. Gemini, with its two-man crew, could go farther and last longer. And now, in 1968, the Apollo flights were poised to land an astronaut on the moon.

On the drawing boards was the design of a beautiful, exciting, complicated, efficient space vehicle called the Shuttle, including its orbiter. This sleek craft—no doubt one of the world's most complex vehicles—was called a shuttle because it was being designed to transport astronauts and cargo into space orbit and return to its launching point. The shuttle design evolved from an earlier concept, the X-20 Dyna-Soar (Dynamic Soaring vehicle) which Randy had helped to design and test in 1962. The most important aspect of the Dyna-Soar project was the testing of the astronauts on the centrifuge in the Dyna-Soar system. Eight military astronauts were tested on the Dyna-Soar cockpit on the human centrifuge. Some of them later became NASA astronauts. Among these was Navy Lt. Dick Truly, who later headed NASA after his space journeys.

Several additional Dyna-Soar centrifuge research programs were conducted for the purpose of training military astronaut candidates from the U.S. Air Force Aerospace Research Pilots' school, the military program which was merged into a civilian program at NASA. The Dyna-Soar X-20 evolved into the shuttle.

The shuttle was not designed to be a moon visitor. Instead, it was intended to explore outer space and to "shuttle" astronauts and supplies back and forth to the proposed space station. It was hoped that eventually the space station would serve as a refueling stop for space vehicles enroute to exploring outer space.

The shuttle was a glide vehicle and had wings and tail fins, giving it some resemblance to an airplane. The Mercury, Gemini and Apollo spacecraft, being drag vehicles, looked clumsy by comparison.

Ten years may seem like a long time to most of us. However, the ten years of intensive space research—from 1958-1968—had produced remarkable progress for such a short period of time. The manned space research projects were conducted like a massive symphony orchestra rehearsal. Scientists of numerous specialties from government labs, universities and industry, combined various parts of the manned space efforts to add—piece by piece—to the data bank essential for successful space exploration.

The task had been much more tedious than anyone had predicted. Unlike other scientific breakthroughs in which a principle, once discovered, could be generally applied, much of the space research data was specific to one vehicle or crew size. The shuttle required different training from that which prepared astronauts for the Mercury, Gemini and Apollo flights. Practically every aspect of piloting this new craft had to be revisited by all of the life scientists and engineers. For a time there was even a question of whether it would take an entire program of special training to prepare pilots to safely fly the shuttle.

One major purpose of the shuttle was to orbit the Earth for sustained periods of time. This presented many life support challenges as well as an entirely new system for docking, maneuvering, and loading and unloading cargo.

Food, rest, waste management, cargo transfer, guidance and control had to be studied all over again.

With the bed rest and other microgravity simulation studies fresh in their minds, and aware of the debilitating effects of confinement on the body, the scientists tested on-board exercise equipment and studied the types of exercise and the schedule for performing them.

Crew size was another important factor to be reviewed. From crews which began with one, then two, and then three, to crews which were now being more than doubled all at once, the need for research had also expanded. Crew compatibility, personality factors, along with the ability to interface with the equipment and the environment took on new importance.

Early designs of the shuttle were tested on the Atlas, Agena and Titan rockets. The pilot control of the Titan rocket was being considered because it had so many command and control improvements over the early rockets.

The shuttle could be boosted into outer space through its launch stages at much lower G-forces. This was much gentler and took much less of a toll on the astronauts than the drag vehicle launches and their reentries. This meant that pressure suits for shuttle astronauts were different from earlier ones and some of their other equipment had to be refined and updated also. The shuttle was designed to be reusable, in contrast with the one-use, drag style spacecraft. The shuttle also landed on dry land like an airplane instead of being dropped into the ocean and then retrieved.

Even the work on the basic simulators—the human centrifuge, the water tanks and the various atmospheric pressure chambers—now had to be painstakingly adapted to this new project. And after the human factors research was done on the shuttle, waiting in line was Skylab, the U.S. space station of the 1970s.

Other aspects of training which had to be redone included training in communications procedures, environmental control systems, flight operations, in-flight activities and experiments, biomedical monitoring and emergency escape training. Flight task requirements, operation of controls, instrument displays and interpretation and monitoring and reporting were all subjects of study.

Also receiving new training programs were the effects of linear acceleration, spins, tumbles, oscillations and vibrations, effects of reduced cabin pressure and pressure suit conditions, noise, lighting and cockpit instrumentation.

The shuttle offered scientists and others new opportunities for surveying the world. Parts of the deserts, forests, oceans and heavens which had never been seen before could now be viewed from a special section of the shuttle.

For the first time in human experience the entire planet Earth could be seen in its entirety but that too requires special training. The whole landscape was spreading and spinning so fast that the observers had to be trained to interpret the sights which appeared and then disappeared in rapid succession. This created the need for more research on eye protection from high intensity flashes and solar radiation and other radiation exposures. These included goggles, lenses and flash blindness protectors.

Microgravity experiments planned for the Space Shuttle included the design of a passageway between the cabin and microgravity science laboratory which could be used when the science experiments were conducted during flight.

Since the shuttle was designed for longer flights, it also had to have more biomedical sensors both in the craft as well as on each astronaut.

One of the most exciting aspects of the shuttle work for Randy was the planned addition of a microgravity laboratory aboard the spacecraft. On Earth, microgravity had to be simulated in water tanks. Now the scientists and astronauts were going to experience microgravity as the real thing. Microgravity research had great industrial applications for medical research, other scientific projects and manufacturing. What might be called an "unintended consequence" the microgravity laboratory aboard the shuttle soon was staffed with a new category of scientists. In a case of "turning the tables," the space scientists who had worked for more than 10 years to put humans into space were now joined by mission specialists who planned to study earthly problems in the space environment!

The crew of the Space Shuttle Orbiter, Mission 51-F (Spacelab 2). Mission Specialists Tony England, Carl Henize, and Story Musgrave; Payload Specialists Loren Action and John-David Bartoe; Commander Gordon Fullerton; and Pilot Roy Bridges.

The microgravity laboratory aboard the shuttle created the need for a new category of astronaut: the science astronaut. In the ensuing flights many kinds of mission specialists joined the pilot-astronauts. Preparing astronauts to be crew members of the shuttle had complexities beyond those of the earlier flights.

Running matching memory and pilot control tasks were important for all flights. However, the shuttle with its requirements for extended time in space, a variety of missions and crew compatibility had the most extensive and comprehensive training requirements yet seen.

Astronaut selection was an ongoing project, beginning in 1959 and continuing with each new phase of development. The original seven Mercury astronauts were all military fighter pilots. The Gemini and Apollo selection process was expanded to include a civilian research test pilot and aerospace engineer named Neil Armstrong. The shuttle astronauts included a mix of military, civilian and experienced astronauts from Mercury, Gemini, Apollo as well as new candidates.

Other highly-trained scientists of different specialties were scheduled to be added to the flight crew as mission specialists. With the shuttle, various on-board research projects were planned. Probably the biggest change was the inclusion of women.

Women were not included in flight crews until the shuttle and the future space station. At first, women were used as mission specialists, not fully-qualified astronauts. Later, some of them were trained to be pilots. And to their credit, none of the men said anything about women drivers—at least not publicly.

Women were assimilated into the space program quite readily and soon were accepted as crew members and flight controllers. Not long ago I accompanied Randall to one of the annual meetings of a scientific society. I began chatting with two young women and we had a long talk about husbands, children and the usual feminine subjects.

It was much later that I learned that one of the women was an astronaut who had actually flown in space and the other a scientist who had worked in research at NASA Houston. Evidently I was the only one who thought this was remarkable. Career women of my generation were mostly teachers and nurses but all that has dramatically changed.

Some of the male astronauts were also mission specialists rather than flight crews. Eventually, astronauts from other countries also joined the crew, giving it an international quality. All of this gave the life scientists numerous research opportunities—and challenges.

The need for research had also expanded. The life scientists had to tackle the different profiles from drag vehicles to glide vehicles; from earlier rockets to more sophisticated ones; and crews which greatly increased in size.

Another aspect of the shuttle program was the increase in interest in the basic sciences. Dr. Wernher Von Braun dedicated his latter years to education, setting up worldwide educational television programs. Children who aspired to become astronauts were directed to the nearest science classes. Interest in biology, physics, chemistry and astronomy soared as the astronauts and mission specialists began communicating with school groups of all ages. The researchers studied the crew size changes in small increments. The first shuttle flight was on the Columbia and had two crew members, Astronauts John Young and Robert Crippen. This was followed by three more Columbia flights with two astronauts each.

The next Columbia flight went into space with a four-man crew. This was followed by the maiden voyage of the Challenger which went aloft with a crew of four. The next Challenger flight had a crew of five, including the first woman, Sally Ride. (No doubt a number of women began hoping against hope that Sally Ride wasn't going to be asked to make the coffee!) After another Challenger flight of 5 crew members, the Columbia flew with a crew of 6.

Then in October, 1984, Columbia flew with a full crew of 7. Probably no one but the scientists realized that so much meticulous research had gone into the expanding crew sized.

A major complaint of space flight critics was the enormous expense involved. A great advantage of the shuttle was that it was designed to make multiple flights. Some scientists and engineers predicted that each vehicle could make 100 flights. Sadly, this did not prove to be the case. Both the Challenger and the Columbia met with catastrophe.

The Challenger blew up during launch on January 28, 1986, and the Columbia burned up on reentry on February 1, 2003. All crew members on both spacecraft were lost.

When the first designs of the shuttle were made public, it was easy to see that with a little imagination one could visualize this new space shuttle as an exotic ocean liner equipped to tour the heavens instead of the oceans.

Just as cruise ships recovered from the Titanic disaster and were redesigned with improved safety features—the shuttle program has become much more safety conscious as future flights are planned. The shuttle is a partner of the space station and both are the subjects of continuing research and development.

The development of the shuttle was the first half of the new series. The other half was the space station. The idea was that the space station would be a stopping place which would offer food, rest and other help to travelers navigating the heavens. As old-fashioned as it sounds, the space station plan sounds a lot like the way station on the Pony Express routes on the western frontier in the 18th century. Even though Skylab was the one space fans of the era remember, there were several space stations on the drawing boards. One that is seldom mentioned was the first—a military station called MOL (Manned Orbiting Laboratory).

The idea for the military manned orbiting lab was to create an Earth-orbiting research station which could identify potential targets as it traveled around the world in an orbital path. This involved superimposing space science on the study of the ability of humans to carry out visual and acoustical surveillance tasks in the ever-changing space environment. It was the same types of human factors variables studied for the early astronauts with ergonomic and flight performance tactics thrown in.

The life scientists and engineers tackled basic questions to add to their knowledge of prolonged work in space stations.

Randy's involvement with this research began at the Navy's aerospace medical research department in Pennsylvania and then continued at NASA Langley. At the Navy's lab the scientists developed several MOL simulations. One facility consisted of a large room, in the centrifuge building, full of electronic and life support equipment. Variables to be studied included habitability, work-rest cycles, effects of prolonged confinement and human capabilities for visual and acoustical monitoring methods and equipment.

Added to these variables were control and display requirements, decision making and information processing skills, and psychomotor, gravity-dependent and weightless simulations. These variables were studied first in the room and then repeated on the centrifuge at a steady rotational rate. The most unusual centrifuge study in this series was conducting centrifuge runs with models of ships and submarines in the visual simulation displays. This would give the test pilots and astronauts practice in identifying "targets of opportunity" during the simulation of the 15,000-18,000 miles per hour orbits. At those speeds the viewing time to identify an object would be measured in milliseconds. Randall continued this work at NASA Langley for Skylab, the civilian space station of the 1970s.

At that time the most unique research project for Skylab—or any other space project for that matter—was the 90-day test. It was not hard for me to

This NASA photograph shows the very successful liftoff of Skylab.

predict that Dr. Chambers would be involved in this unusual project. It was preceded by a 60-day test in which he was also involved.

Staffing an orbiting space station would require crew members who could perform their tasks for long periods of time, working in the unusual circumstance of being confined and traveling at 17,000 miles an hour—all with portable life support systems, including imported atmospheric pressure. This work was done in a number of places by scientists and engineers from government, university and industrial labs. Dr. Chambers was in charge of the human factors and life support portions of the study.

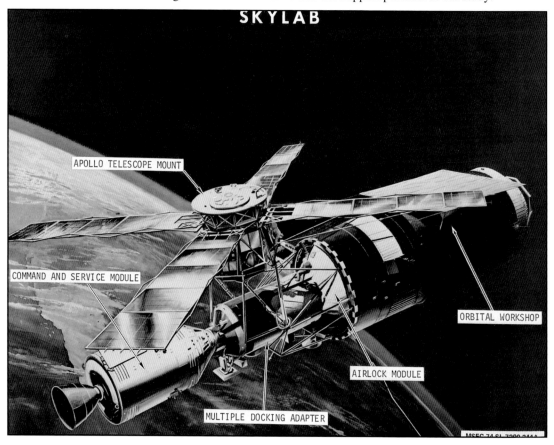

This image is an artist's concept of the Skylab in orbit with callouts of its major components. In an early effort to extend the use of Apollo for further applications, NASA established the Apollo Applications Program (AAP) in August of 1965. The AAP was to include long duration Earth orbital missions during which astronauts would carry out scientific, technological, and engineering experiments in space by utilizing modified Saturn launch vehicles and the Apollo spacecraft.

The study involved confining four healthy young graduate students in a closed facility for 90 days. They were to be confined in a double-walled steel cylinder that measured 40 x 12 feet and would be home for the four subjects for 90 days. Nothing was to go in or out during the 90 days. Food, water, and other supplies were stocked in advance. Bodily wastes were to be treated by special equipment and then reused. There were numerous voice and video communications which enabled scientists to study and communicate with the subjects. The subjects could also communicate by telephone with family and friends. (They probably learned later that their conversation was recorded and studied as part of the project!)

After all of the supplies were loaded and all of the pretest arrangements made, the doors were locked and sealed.

Activities similar to those needed to operate a space station were scheduled. This included work cycles, reconnaissance tasks, flight control operations as well as preparing their meals, doing prescribed exercises on a bicycle ergometer and learning to perform personal hygiene tasks, as well as maintaining work-sleep cycles. A day in the life of a 90-day test participant had a list of 18 "to do" items to be performed. These consisted of schedules for eating, sleeping, working, exercising, cleaning up and taking a turn at equipment maintenance. There was also a period each afternoon for leisure activities such as reading, board games or television. All of this required enormous coordination, cooperation and communication among the four subjects as well as between the subjects and the science and medical monitors outside. Monitoring was conducted 24 hours a day.

One of the biggest challenges was the maintenance of crew compatibility while encased in integrated life support systems for 90 days. Eight young men were chosen as semi-finalists from the group of applicants who wanted to participate in the study. All eight completed the pretest training, which was extensive. Then the four finalists were chosen after all eight had passed numerous selection tests. They also promised to stay confined for the total duration of the test and not to ask to come out under any circumstance. The purpose of the project was to develop information and design requirements for a closed-cycle life support system for crews in space. The reasoning was that if this could not be done successfully on the ground, the whole idea of selecting crews for a real space station was problematic.

All went well for two months and then—perhaps understandably—the little habits of each man began getting on the nerves of his companions. In a scene familiar to anyone who has ever gotten annoyed with a roommate, a verbal confrontation came on the 56th day. As inconsequential as it sounds, the argument was precipitated by the refusal of the crew commander to allow the others to sit in his chair. There followed a litany of pent-up aggravations about the annoying habits which would have been petty under normal circumstances.

Such complaints as, "Do you have to brush your teeth that way?" Or, "Do you have to use so much soap and water to bathe?" Or, "Can't you clean the sink after you shave?" These small complaints became very important in such a confined space. With the study in jeopardy, the scientists had to restore order by remote control.

The subjects then organized a video conference in which they vented their feelings. In typical fashion, the scientists recorded and analyzed the proceedings. In post-test psychological exams, it was discovered that three of the four subjects had actually undergone personality changes during their confinement.

Their psychomotor skill level as shown by performance on the Langley complex coordinator was found to be unchanged. This suggested that the interpersonal relationships were more important to the success of the mission than other abilities. After the 90-day period, the subjects were tested extensively for an extended period of post-test study. The purpose was to develop information and design requirements for closed-cycle life support systems for crews in space. The data also became valuable to the Navy and was used in the study of crew compatibility among submarine crews that also spent long days of isolation and confinement at the bottom of the ocean.

Much of the data was incorporated in the staff and operation of the real Skylab as it was designed and built. Skylab—a working, orbiting space station—was a thing of beauty to the scientists. One of its outstanding characteristics was its "ship shape" condition, everything stowed away neatly in its place. This neatness distinguished Skylab from other large spacecraft and can probably credit its Navy origins for this distinction.

Many scientists—including Dr. Chambers—still speak with regret that Skylab was not maintained and allowed to continue its historical flight profiles. They will tell you, to this day, that the scientific studies accomplished in Skylab were very worthwhile, indeed.

At a space seminar in 2003 Randy discovered that the cold war competition between the United States and Russia is not a thing of the past. The meeting was part of an aviation festival and was held in a large tent on an Air Force Base. The seminar speaker was a Russian Cosmonaut who did his best to speak over a severe thunder storm. Finally, the power failed and the meeting place was plunged into darkness.

Kansans—who live in tornado alley—accepted the power failure as nothing unusual—but the Russian visitor seemingly could not believe that a modern Air Force Base could not provide a "back-up" power supply. He turned to the audience and said grandly, "That's why the Russians are providing the brains for the International Space Station and the U.S. is providing the money!"

Skylab 3 On-Board Photo – This outstanding view of the Skylab space station cluster in Earth orbit was taken from the Skylab 3 Command / Service Module during the "fly around" inspection by the CSM. Clouds and water are below. Note that one of the two solar array system (SAS) wings on the Orbital Workshop is missing.

This Skylab 2 crew portrait portrays the mission's crew of three astronauts (left to right): Joseph P. Kerwin, science pilot; Charles "Pete" Conrad, Jr., commander; and Paul J. Weitz, pilot. This crew made urgent repair work on the damaged Skylab to make it operational and habitable.

In the late 1950s the general public, absorbed with the affairs and events of this planet, paid scarce attention to the Sputnik launchings. Scientists of every type were amazed and challenged by such a complicated feat. Their fellow citizens were much less impressed. (Probably because they were totally uninformed of the challenges presented by the effort to get off the planet.) Several years later, however, when astronaut training began among great furore and extravagant publicity, the astronauts became instant celebrities.

Space scientists were also thrust into the limelight—a place which made these serious, bookish introverts as uncomfortable as if someone had asked them to sing or to perform magic tricks. It would probably not be an exaggeration to say that they became "accidental" celebrities.

It took the press a while to find—or train—journalists with even an iota of scientific knowledge. Indeed, the first journalists sent to interview the scientists at space research laboratories were general assignment reporters who reflected the failure of the nation's schools to teach much science and engineering. The general public was divided on the subject of space. Some people were enthusiastic about President Kennedy's mandate to put a man on the moon within the decade. In this group were the space zealots who had been waiting for years for such an opportunity to present itself.

Others thought the whole idea was nonsense—the most ridiculous thing the politicians had ever come up with.

Then, of course, there were the humanitarians who felt that money could be better spent on helping the poor—even though money seldom trickled down to that group. And many of the space exploration critics rarely contributed much of their own money to charity when they had an opportunity.

Our first challenges were with our neighbors. We had moved into a pleasant, picture-book neighborhood where lawns and shrubs were well-trimmed, dogs were properly leashed, and traditional values were not only respected, but expected. Mavericks in the suburbs in the 1950s were few and far between. Outwardly, it looked as though we had the proper credentials to fit right in. Our differences with the neighbors were subtle but marked nevertheless. Here was a young doctor who evidently had never healed anybody of anything and didn't appear to be preparing to start. Instead, he seemed to be working long hours at the nearby naval air station, doing who knows what?

We looked like an average family next door when we moved to Hatboro, Pennsylvania in 1958. By the time this picture was taken, about three years later, the neighbors had discovered that we were really the ground crew and unofficial headquarters for space experimentalists.

Reports from there were that Dr. Chambers headed up a group of wild experimenters who rode the human centrifuge until they blacked out and then began submerging each other in tanks of water for long periods.

To point out that these experimenters were figuring out the best way to train astronauts to travel to the moon did not enhance their reputation. Added to this scene was the fact that a variety of space visitors from all parts of the country seemed to gravitate to our house.

Deborah Franklin, the wife of Benjamin Franklin who was studying electricity by experimenting with kites during thunderstorms in the 18th century, summed up my situation as well. Evidently, the scientific-type man has not evolved much since then.

Mrs. Franklin is quoted as having said, "Every nut who comes to Philadelphia makes a beeline for our house."

After living among the scientists for a while, I learned that they think—and work—differently from other people. They have what I call the "all-out personality." This means that anything you think is worth doing is not only worth doing well, it is worth going all-out to the point of overkill.

It also means becoming oblivious to the mundane things of everyday living. Our neighbors were dedicated to eradicating crab grass before it had time to take root. Their space scientist neighbor rarely even looked at the grass and took the attitude that growing things in our yard were on their own. Ours was also a neighborhood of neat, well-kept houses and Randy's study soon became a legend. It was filled to overflowing with books, journals, notes, maps, graphs and charts along with being the headquarters for Fred, the skeleton; the pickled brain made famous at Show and Tell; plastic models of the human eye in a greatly enlarged form and odd pieces of flight suits being studied and tested. All in all, his collection of oddments rivaled anything outside of a museum. One might say that his study was a rat's nest that even a rat would try to tidy up. I gave up being embarrassed about this when I realized that the visitors whom he invited in felt right at home there and viewed the chaos with great appreciation.

Another aspect of the all-out personality is that dedicated scientists pay no attention to what anyone thinks of their projects. The question, "What will the neighbors say?" rarely even comes up. Randy seemed unaware that people might wonder about someone who spends the night in a water tank, or displaces a dolphin who is peacefully swimming in a public aquarium, or who competes with a chimpanzee to see who can learn to operate the controls of a spacecraft faster. To him, it was all in a day's work.

Our sons also helped to contribute to the idea that our family had an odd life style. When Craig's class was studying the senses, he asked to take the pickled laboratory human brain to school. The brain is sealed in a round plastic container and is one of Randy's prized accessories.

I hesitated to do this because I was afraid it would cause a disturbance. However, several grades filed past the brain with the dignity of medical researchers, and the teachers thanked me for my contribution to education. I was beginning to feel rather good about the whole thing. Then the scholastic mood was shattered completely. Craig's class went to the library and a friend of his could stand the suspense no longer.

He called out in a very loud voice, "Craig, whose brain is that?"

And Craig ruined the academic atmosphere for the rest of the day by answering, "It's my father's!"

Mark built a miniature centrifuge for a science fair project which gained a lot of attention and awards. It consisted of several clear plastic tubes which contained colored balls. When activated, the arms spun around lifting the balls to the top of the tubes, illustrating the phenomenon of centrifugal forces. In a neighborhood where conventional children played with conventional toys, our children opted instead to imitate their elders. Several people summed this up with the old saying, "The apple doesn't fall far from the tree!"

One day I just happened to look out a window in time to get a glimpse of Mark and a neighbor boy dragging a long ladder to the back of the house. I followed them just in time to interrupt their plans to jump off the roof with umbrellas. Mark and Craig listened intently when some of our visitors discussed a research project which called for the pilot to jump out of an airplane encased in protective foam instead of being attached to a parachute. This prompted the boys to wrap themselves in cardboard

(Right) When their father began training the early astronauts, Mark soon realized that he had a treasure trove for "Show and Tell." Fred the skeleton, the pickled human brain in a plastic case and the enlarged plastic model of the human eye were used by their father to study the limitations and capabilities of the human body; Mark borrowed them to dazzle his classmates.

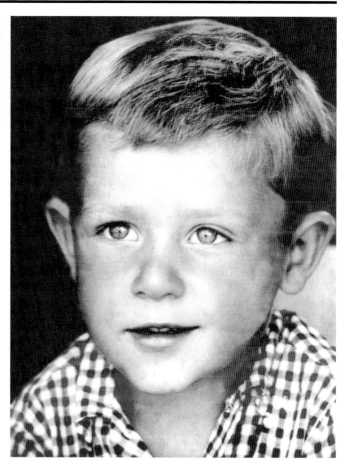

boxes and throw themselves down the stairs. (I think the adults' plans failed to materialize and I never found out if anybody had actually tried jumping out of an airplane encased in foam.)

From time to time our house was filled with assorted experiments the boys were trying out in zoology, botany, biology, psychology, chemistry, and various kinds of engineering. One of our neighbors in Pennsylvania probably never forgot one of the Chambers' boys experiments. The boys were trying to create a scientifically-correct environment for two turtles about the size of small dinner plates. They built a fish pond in the back yard which looked like a sunken Roman bath tub. The turtles, which had come from the local pet shop, went wild when they discovered a nearby creek and they frequently ran away from home. The boys were often out trying to retrieve them.

One of our neighbors had a tendency to over-imbibe and one afternoon he had just awakened from a bender when my boys knocked on his door and asked, "Did you see a big turtle go by?" Evidently his imagination ran away with him and he visualized a much bigger turtle than the ones that were lost. And he wasn't really sure whether he had seen them or not. He took a pledge of abstinence that lasted longer than any of the previous ones.

After we moved to Virginia both boys were active in many science fairs and competitions.

(Right) Little brother Craig followed in the family tradition when he got to kindergarten.

One evening early in the space program Randy invited our family to go on a little ride. We went to the flight line at the Navy base and waited for a small plane to land. Emerging from the plane were Alan Shepard and Scott Carpenter. Randy had arranged to drive them to a motel, unobserved by the press and other fans.

At the motel we waited while they tried to untangle errors in their reservations. For awhile it appeared overly complicated. Our son, Craig, who was about three at the time, walked up to Al Shepard and said solemnly, "You can have my bed!"

The child evidently realized that these were two American heroes.

After Alan Shepard's historic Mercury flight, the press came knocking on the laboratory doors. There was puzzlement all around. The reporters didn't know what questions to ask and Randy and the others didn't know what to tell them about the Mercury project. There were two main problems—one was that the public's lack of scientific knowledge was prevalent and the other was the government's restrictions on information. It was hard to know which group was more ill-at-ease—the reporters who were trying to find a story or the scientists who just wanted to get back to work.

After that the reporters descended on the laboratory quite often. Several times they even asked for copies of the scientific reports Randy was going to present to some conference. Usually, that was the end of it.

The usual bureaucrats that run the government were also puzzled by the workings of the scientists. This became apparent when an efficiency expert arrived to study Randy's activities in compliance with a governmental decree.

The man evidently tried taking notes and following Randy around among the various research sites. However, he lasted only until noon. He went out to lunch and never returned, leaving questions as to whether he had left the entire profession or just the simulation laboratory. Even the well-planned "photo ops" didn't always work out the way the planners had hoped. One of these ill-fated events called for Dr. Chambers to sit in a simulator with a visiting beauty queen. He was supposed to be photographed explaining the control panel of a spacecraft to her. However, the situation was so awkward that the woman looked like a student trying desperately to understand algebra as it was being explained by a professor on his way to death row.

Television, which was still in its infancy, had many of the same problems as the print media, except theirs were aired in front of cameras. After the first Mercury flight, a television reporter thrust a microphone under Randy's nose and asked, "What did you think of Alan Shepard's flight today, Dr. Chambers?"

The reporter acted as though he expected some memorable quotation worthy of a rocket scientist.

Randy, completely at a loss for simple, non-technical words, said, "I think it was just great!"

Several years later the public relations venue was a museum. The idea was that Randy would accompany the television reporter around a museum. She would point to an object and he was supposed to explain what it had to do with the space program. She was a very pretty woman, which persuaded me that print journalism had been a good choice for me. The problem with this assignment, however, was that she hadn't done her homework. The first object she pointed to was an airplane that had little, if anything, to do with the space program. This caused her subject to give some kind of vague pained response. The next item was a display of a life support system which for a moment looked like it might rescue the interview. "Well, that's a life support system," Randy told her.

Then she stunned everyone within hearing distance by asking, "What's a life support system?"

I think they went to a commercial right after that.

During the early astronaut training days Gus Grissom occasionally called Randy on the telephone. It was quite startling to me the first time I answered and he asked, "Is Randy there? This is Gus!"

He didn't seem to feel the need to identify himself further. The calls were part of his direct and straight forward approach to everything. He liked to make his own arrangements and this was usually why he called. He and Randy also reminisced about their boyhood days in neighboring Indiana communities.

Years later we visited the Gus Grissom Museum in Mitchell, Indiana. Randy told a skeptical guide that he had helped train Gus.

To demonstrate this, Randy suggested the guide reach inside the helmet on display. He told her that there was a tiny microphone inside the helmet. To the guide's utter amazement, the microphone was there. (This incident taught both the guide and me that the helmet was more than a helmet—it was a miniature work station.)

In this publicity-mad world it's probably very puzzling to the media to meet scientists who surround themselves with journals and textbooks instead of comics and sports pages. For their part, trying to describe what they're attempting to do in a few sound-bite segments is thought to be impossible. Over the years the press has descended on Randy whenever there is some major space event in the news. The reporters now seem more knowledgeable, but most of them are like the efficiency expert of long ago—they can't really figure out what a scientist does. And accuracy still eludes them. Not long ago reporters from a local television channel interviewed Randy for at least two hours. Then they introduced him on the air as "Richard."

During the early flights Randy also was invited to appear on one of those late-night radio talk shows. I went with him to the studio and soon regretted it. For one thing, I learned that there is a population of strange characters—most of them probably insomniacs—who evidently listen to the radio all night. One after the other these night owls called in with questions and ideas about everything from pre-historic man to UFOs which they claimed had visited them personally. They also began seeking verification for some physics formulas or some other highly-technical plan which they alone understood. One real space zealot talked about the creatures that he is convinced already inhabit Mars. As the time grew later I began to realize that I wasn't even concerned about whether creatures lived on other planets. I just wanted to go find a bed on this one!

If anybody had asked for my opinion (which is unlikely) I would have suggested that all of these people get more sleep. In the early days of space research, scientists still enjoyed their annual meetings. They got together in small and large groups to enthusiastically share their exciting research projects. Now in the age of computers and all manner of electronic communication devices, these meetings are less enjoyable. Scientists are much less likely to share their data and they spend more time in solitary effort at a computer. A lot of the previous enthusiasm has been replaced by a communications overkill.

One memorable meeting during the Apollo flights was an annual meeting of the American Association for the Advancement of Science in Boston. The scientists were especially enthused because the centerpiece of the discussion was a collection of moon rocks which had recently been brought back by the Apollo astronauts. The rocks were of great interest to scientists of many disciplines.

It soon became apparent that not all of those gathered at the convention hotel were space fans. A rather large demonstration protesting money spent on space exploration to the neglect of the domestic need was loudly in progress. The demonstrators paced in front of the hotel carrying signs with sayings such as "The poor can't eat rocks," or "Moon rocks cost more than diamonds."

Obviously, this kind of publicity is not good for any cause and the usual response to protesters is to try to get rid of them as quickly and as quietly as possible. In this case, however, there were behavioral scientists at the meeting. They invited the protesters in, gave them chairs and even gave them a chance to voice their opinion. I doubt that the scientists made any converts; however, the situation was soon quieted without undue negative publicity. It was probably the middle of the following week before the protesters realized that they had been hoodwinked by some of the best behavioral scientists in the profession.

Another memorable meeting took place in the early 1960s in San Juan, Puerto Rico. Randy was to be one of the keynote speakers and I was invited to come along. It was memorable to me because this was my first airplane flight over the ocean.

One would think that having an aerospace expert as a traveling companion would calm one's fears. I discovered that just the opposite is true. In the preliminary safety talk the stewardess showed us how we were supposed to use the seat cushion as a flotation device if we had to abandon the airplane in the ocean. I don't know which scientists worked that out but it sounded like a dumb idea to me. "Who has the parachutes?" I asked Randy.

"Oh, there are no parachutes," he replied matter-of-factly, "they just wouldn't be practical in a commercial jet."

"Not practical?" I repeated. The ocean was five miles straight down and he was telling me that a seat cushion—questionable at best—was more practical than a parachute?

This logic still escapes me. I know how to swim, but I don't know how to fly!

"What keeps this airplane up here, anyway?" I asked, somewhat belatedly.

"Oh, we're riding on a cushion of air," he said. "Air?" I sputtered.

He sat back in his seat and said these reassuring words, "Don't worry, dear. If something happens to this airplane, you'll never know the difference."

There were more hair-raising moments to come. The taxi driver who was assigned to drive us from the airport to the hotel was a very friendly fellow. He shook hands and talked to us the entire trip while facing the back seat. I never learned if he had some kind of radar that enabled him to narrowly dodge objects in front of him while he was glad-handing to the people in the back seat.

At the hotel where the speech was to take place I was given an orchid the size of a dinner plate—and then whisked away by a hostess who was obviously assigned to entertain me far from the meeting. (This evidently had something to do with their cultural attitudes about a woman's place.) Randy reported later that there were more nerve-wracking moments at the meeting.

The committee chairman had decided that Randy could not speak until they had placed an American flag by the podium. (I never learned whether this was considered courtesy to a visitor or some sort of political requirement for a government representative.) Anyway, they dispatched an aide to go and pick up a flag and delayed the meeting until his return. He, in turn, delayed the meeting several hours because he stopped to visit his sick cousin enroute to returning to the hotel with the flag. On our return flight to Philadelphia I kept wondering if the airplane maintenance had been performed by someone like the flag-bearer!

For several years during the 1960s Dr. Chambers participated in meetings of the Aerospace Medical Association, the National Academy of Sciences, the American Institute of Aeronautics and Astronautics, The Human Factors Society, the British Interplanetary Society, the American Psychological Association, several NATO conferences in Italy, Greece, and Germany, and the American Rocket Society. He was also kept busy with engineering and consulting advisory groups in NASA and the Department of Defense.

As the space research progressed, the public awareness increased. Before long, Randy was much in demand as a speaker for church and community groups. The typical pattern was to invite him to speak at one of the group's monthly meetings which usually began with a dinner. (The standard fare was chicken and mashed potatoes which seemed to be some sort of basic requirement.) The audience had meager and varied knowledge of science and often looked at Randy like he was speaking a foreign language. More importantly, the scientist's remarks had to be approved in advance by government officials. Even though we live in a democracy, the government is still a bureaucracy.

A government employee was considered to be government property no matter where he went or what he was doing. He was always reminded that as a representative of the space program, he had to be careful not to provide ammunition to the space critics. The press was especially suspect because they often roamed around meetings looking for a story and exaggerated the incidents of injury or risky experiments they discovered.

The irony of all this was that there is probably no one around who is less of a rabble-rouser than Randall Chambers. He is a space fan through and through and has diplomatic skills that rival real ambassadors.

The local press was probably more of a problem to him than the program chairmen. On one such occasion a reporter demanded a copy of Randy's speech before the meeting.

"How can you quote my speech before I have given it?" Randy asked.

Another hapless reporter arrived with an interview already prepared. He evidently had done some library research and thought he knew what Randy was going to say. Even though I was usually sympathetic to members of my profession, I had to admit that the pre-written interview was outrageous.

Space scientists were also much in demand at schools and were invited to speak to classes of all ages. It was not hard to see that the space program had greatly increased the interest in the study of science, especially physics, astronomy, biochemistry and biology.

One day that stands out in my memory was a day in which both of us were invited to participate in a "guest speaker" day at an elementary school. I was assigned to a second grade class where I proceeded to try to explain to the class what a journalist does. This effort could not be called a total success. The second graders were cute as could be, but most of them were still struggling with reading, much less writing. The big hit of the day was the space pioneer assigned to the fifth grade. Those children peppered Randy with all kinds of questions and even extended his visit by voting to skip recess! Randy, however, met his match on another school visit, this time to first and second grades. He had a miniature model of a Mercury capsule which he held up for the class to see.

"You can see," he said, "the Mercury capsule for John Glenn's flight was much smaller than later spacecraft."

No sooner had the words left his lips than both of us realized that there was a great misunderstanding. The children took his remarks literally and were greatly puzzled as to how an astronaut could ever cram himself into that little model!

In 1969 I became a sort of "second class" celebrity. My first book, *Don't Launch Him—He's Mine!* which was a humorous account of my life among the scientists, was condensed in Reader's Digest to coincide with Neil Armstrong's walk on the moon. Then I began receiving invitations to speak at meetings and on television.

I was invited to speak to many community groups and soon discovered that I had different problems than Randy. I had no government officials monitoring my remarks, but I was hampered by having such a superficial knowledge of science. However, the women's groups to which I spoke enjoyed my stories about Randy's study and of my "glorious incompatibility" version of marriage.

One woman even suggested that I write a sequel to Don't Launch Him entitled, "Launch Him—I've Had Him!"

I also discovered that television appearances are not for amateurs—at least not this one. Being on the other side of the camera is a daunting experience. My first experience was on a local television station in Norfolk. One of my neighbors had driven to the station with me for moral support. I didn't realize I was going to be so nervous until the time came to introduce my neighbor.

"This is my next of kin!" I babbled.

This experience scarcely prepared me for an appearance on a national talk show. I discovered that the staff of these television shows do their best. They had a makeup artist who transformed my face and a hairdresser who did marvelous things to my hair. They put so much makeup on me that when I passed a mirror I looked at my reflection and wondered who that woman was.

Then they opened the stage door, gave me a slight shove, and there I was with all of those bright lights shining in my eyes and six million people staring at me. And instead of being able to discuss my life among the space scientists as I had planned, I found it hard to remember my own name. The only experience I have had previously which compares with going on television is going to surgery. And if somebody gave me a choice between the two right now, I'd probably ask, "What kind of surgery do you have in mind?"

Another of my invitations was to help celebrate the 100th birthday of the Norfolk public library. I enjoyed playing my role as a writer because libraries really seem to appreciate authors. Then I overheard two young girls discussing whether to ask for my autograph.

"I think we should get it," said one, "writers live in libraries, don't you know that?"

And then she ruined the 100th anniversary celebration for me further by adding, "I bet this one has been here all the time!"

I also became accustomed to less-than-flattering remarks from adults. After giving a number of presentations I decided that someone should prepare a manual to help prevent cruel and unusual treatment for guest speakers. Several times I arrived at a meeting place to present a program that they had publicized widely. However, when I told the person at the door my name, she said with a casual wave of her hand, "Just sit anywhere!" Another time a woman greeted me by saying, "You picked a good day to visit us. I hear we're going to have a great speaker." I said, "I hope so!" No matter how hard the speaker tries to make the arrangements go smoothly, the program chairman seldom follows through. Ask for a microphone and nobody sets it up and tests it until after they introduce the speaker. And then the committee spends the next 10 minutes trying to get the microphone to work properly.

After I gave one of my talks to a community group, the program chairman thanked me profusely. Then she added, "Next month we will have a real celebrity!"

During a question-and-answer period a man asked, "Doesn't it bother you to be around such big brains all the time?" Another one of my "fans" asked, in all seriousness, "Are you an author or an authoress?"

People born after 1950 probably take the space age for granted. These days the public is much better acquainted with space projects and the critics much less strident. Most people would have a difficult time imagining those long-ago days when nobody even considered getting off the planet, much less wondered how to go about it.

The skeptics, critics and outright opponents of the space program evidently were unconvinced regardless of the progress the program had made. When Neil Armstrong and Buzz Aldrin walked on the moon there were many people who claimed that the whole thing had been staged—that it wasn't possible to travel to the moon. The most amazing thing to me was that there was a large group of these skeptics in the Langley, Virginia area, many of whom frequently passed by the Lunar Lander simulator!

Randy also encountered skepticism among some of the Indiana farmers he encountered. One of these skeptics challenged him by saying, "You ought to be ashamed of yourself—trying to get to heaven before God wants you to go there!"

Randy gave him a calm, patient answer: "If God doesn't want us to go into space we'll never get off the ground!"

Serendipities—those good things discovered unexpectedly—abound in any account of space research.

For human factors scientists and their dedication to man-machine systems, the dawning of the space age in the mid-1950s ushered in the most productive period of their emerging specialty. Their scientific discipline was a newly-forged combination of several fields including aeronautics and astronautics, engineering psychology, human engineering, physics, astrophysics, biophysics, biochemistry, human factors psychology, human-machine interface, aerospace medicine and health, aerospace physiology, industrial psychology, industrial engineering and ergonomics.

Space research was made to order for this discipline and this new discipline was made to order for it because manned space flight involved such extraordinary coordination between the human astronaut, his protective clothing, his biomedical monitoring, his life support system, his communications system, his spacecraft and rocket. A space flight also demanded coordination between humans in space and those on the ground. No one scientific or engineering discipline could have prepared the astronauts for such a venture.

The sheer complexity of the requirements for putting humans into space and returning them safely at first seemed almost insurmountable. They would have to be equipped and prepared to function under extreme conditions never before encountered by humans. As it turns out, preparing and protecting astronauts from such extreme conditions brought numerous serendipities which help to improve the lives of the earthbound.

It seems that equipment designed and developed to be tough enough for use in space can be useful when adapted to use on the planet Earth.

The dawning of the space age in the mid-1950s ushered in a period of intensive research. There was much work to be done: outfitting an astronaut with protective clothing, equipping him with a fail-safe life support system and biomedical monitoring, reliable communications and then launching him into space in a vehicle sturdy enough to withstand the rigors of intense heat, sudden cold, speeds heretofore untested and a variety of atmospheric and weightless environments.

There were few ready made items that could be ordered from catalogs and few training manuals available.

The most significant "side effect" of space research is probably the least known. It is the establishment of principles and standards of human capabilities and limitations.

Before the era of manned space research, machines of all kinds were presented to the human operator who was then challenged to learn to operate them. Space research extended this process as it worked on the idea that machines and other equipment be made to meet the pilot's capabilities.

Up until the beginning of manned space research, capabilities and limitations of humankind were largely estimates. However, intensive research and testing to prepare humans to work and survive in the harsh environment of space helped establish the criteria for physiological and psychological capabilities and limitations.

Space vehicles—with their accompanying support systems—are among the most complex inventions ever produced. The space environment is hostile and unwelcoming. This pushed human capabilities to the edge. This meant that if the astronaut were to succeed in traveling into space and returning safely, his equipment had to be as well designed, as well-made and as well-placed as possible. And the idea of making equipment "User Friendly" was born! Gradually, over the ensuing years, standards, guidelines and mental workload assessments established in space research have been utilized in various ways for the betterment of life on this planet for millions of its inhabitants.

It has been an example of an enormous technology transfer.

An important utilization of the guidelines and standards has been in setting safety requirements for the manufacture of tires, brakes, windshield glass, mirrors and control panels for cars and trucks. Also lighting for both cars and highways, roadside signs and warning signals of all kinds. The guidelines developed for astronaut performance have been transferred to work-rest guidelines for airplane pilots, truck drivers and bus drivers.

The principles of training to-do tasks such as operating ground-based equipment have been borrowed from the intensive program of learning, conditioning, familiarization and accommodation required in astronaut training. Advanced muscular conditioning devices and procedures were provided to athletes. The restraint systems designed for astronauts has inspired great improvements for restraint systems for cars, airplanes and child seats.

The mannequins used to test the safety of cars and other equipment had their origins in crash dynamics programs. Much of astronaut training was devoted to planning and practicing emergency procedures for escape from dangerous situations such as vehicular failure. This was necessary because there are so many things that could go wrong in spacecraft and in outer space and planetary environments. Many of these same techniques have been utilized in updating plans and practice sessions for emergency escape from earthly disasters such as fires, floods, earthquakes, plane crashes and other catastrophes.

A great deal of research and testing went into protective clothing and other equipment for the astronauts and the dangerous environment they faced. Many of these innovations have been incorporated into special clothing for earthly workers. Fire retardant clothing developed for astronauts is now being used by firemen and police forces. Also, protective clothing designed for space travelers is now used by outdoorsmen, sportsmen, sailors, submariners and security forces. Astronauts were outfitted with a variety of helmets which were sturdier and more comfortable, with numerous special features. These are now being utilized in the development of helmets for bikers, divers, sportsmen and others.

Thanks to space research, Velcro, zippers and goggles are now more functional and reliable. Also, the shockproof, waterproof watches and high performance writing instruments owe their quality to space research.

Space researchers borrowed heavily from aviation—and aviation borrowed ideas and equipment from space research. Air traffic control equipment and techniques were borrowed from mission control. Returning the favor, mission control borrowed from air traffic control in the development of flight control systems. Metals, alloys, composites, ceramics and polymers that withstand very high temperatures are now being used for welding, aircraft and cars. They were also used for corrosive and high temperature surfaces.

Aerospace technology has been applied to ground-based safety systems including helicopters, buses, trains and ships. Another very important, but seldom mentioned, transportation problem is the maintenance and care of external fuel tanks. Most of us rarely worry about fuel tank maintenance but they have to be protected against extreme temperature, vibration, and noise. The materials and techniques developed for fuel tanks in space vehicles have been widely adapted for earthly vehicles as well as for other vulnerable fuel tanks.

Ventilation and temperature control for space capsules presented a big challenge and much effort was put into producing livable spacecraft. Now this same technology is being used in all kinds of portable heaters and fans for use by earthlings. Space research involved extensive study in finding or developing materials for space vehicles and equipment. Those developments have now been utilized to make earthbound equipment sturdy and reliable.

The space flights virtually revolutionized surveillance technology for satellite monitoring of the Earth on a worldwide basis. Before the advent of space flights the Earth could not be viewed in totality. The influence of technology transfer from space research to medicine and health has been enormous.

Human beings were being sent, for the first time, into an environment full of perils and unknown hazards so biomedical monitoring became a necessity. However, the devices for monitoring blood pressure, temperature, and oxygen levels as well as breathing were very clumsy, hard to set up and awkward to use. It was crucial to be able to monitor vital signs during a space flight so the scientists and physicians embarked on an intensive effort to develop biomedical monitoring that would travel well and was easy to handle.

Since weight was very important in a spacecraft, biomedical monitoring devices also had to be small, portable and lightweight. This resulted in miniaturization of this equipment which is the basis for much of the biomedical monitoring and emergency intervention technology in modern hospitals today.

The transmission of medical training and treatment by remote control from major medical centers to physicians in other parts of the world has been made possible by satellite technologies. Inflight medical support systems are now available for ambulances and rescue vehicles developed first in space research. The value of physiological conditioning and exercise is directly derived from the bed rest studies during space research of how best to prepare an astronaut to cope with the isolation, confinement and weightlessness of space travel. Before the 1960s bed rest was thought to be beneficial and was highly prescribed for a multitude of ailments; the weightless studies which showed the muscular deterioration effects of inactivity taught the physicians otherwise.

Extensive space research on the capabilities and limitations of human sensory organs has led to improved rehabilitation equipment for the disabled. Wheelchairs, custom made for the user, often utilize findings of space research. Also, the equipment which enables persons of low vision to see better, those with poor hearing to hear better, and those with impaired communication abilities to convey their feelings and wishes, all have beginnings in space research.

Space research involved an extensive study of the balance mechanism in the inner ears. These findings have been used in programs to prevent falls and also in sports medicine. Findings from space research are also being used to combat all kinds of motion sickness.

Vision—especially in extremely bright environments—prompted the development of special lenses and visors to help protect the human eye. Many of these findings were adapted to improve surgery for cataracts, near-sightedness and astigmatism. Health-monitoring equipment such as imaging technology, bicycle ergometers, metabolic analyzers, automatic thermometers and blood pressure measuring systems had refinements in space technology. The technology for measuring brain function capabilities during sleep and wakefulness, day-night cycles and work overload were also adapted from space research.

The atmospheric pressure factor, so prevalent in space research, required many hours of study in atmospheric pressure chambers. Many of these findings have gone into the baric chambers now used in hospitals to treat patients suffering from carbon monoxide poisoning and similar conditions.

Biological and chemical research projects conducted in the microgravity laboratory on the shuttle have resulted in the development of pharmaceuticals developed for protection against possible contaminants. Waste management technology and reusable products have also resulted from this research.

Communications were greatly improved by the need to keep in touch with an astronaut. Solar power for ground based systems was borrowed from space research. Communications research includes improved remote sensor technology and communications satellites. Video phones, taken for granted these days, were developed for specific space missions. Night vision technology—used widely by modern military forces—was developed to enable astronauts to view the space environment electronically. Virtual reality technology, widely used on the modern-day battlefield, is also useful in medicine and flight simulation.

Space research can also take credit for perfecting microwave ovens and other food preparation utensils and devices such as Teflon-coated roasting pans and skillets.

Infrared lasers and special cameras and imaging devices now used in medicine and security systems were developed for use in space.

Other space discoveries which are being used by the military include survival techniques, navigational methods, field rations and eye protection. Improved weather forecasting by satellite technology, evolved from the space program, has contributed to meteorology and its application to oceanography and geology.

Space research has contributed to the science of astronomy by increasing the understanding of astronomical objects and events. This includes the special field of radio astronomy and infrared astronomy, which involves sending radio waves and infrared signals to distant planets to measure distance, time and location.

Computer technology has been advanced by space research, especially in the areas of data acquisition and analysis. Also, personal computer-based training systems for use in libraries, classrooms and industry had their beginnings in space research. The field of social behavior has also been impacted by space research. Personnel selection techniques developed for astronaut selection have been validated for selecting personnel in other occupations. The studies of social behavior in the crowded confines of spacecraft used to select and train astronauts are now being applied to select and train workers in other kinds of confined environments such as sewers, buses, submarines and ships.

Space flight simulation has contributed to the environmental simulation technologies in other confined or hostile environments. Ergonomics—the study of work and the work place—has adapted much of the technology developed for the difficult space environment. This includes technology for remote control of buildings, nuclear power plants and manufacturing systems. It also includes working with more efficiency and using ergonomically-designed tools and improved manufacturing technology. This has been especially true in the study of operating systems in unusual work environments such as coal mines and power plants.

Data from biosatellites and biological experiments in space have contributed valuable data to the field of biology.

Interest in the sciences and improved science curricula have been sparked by the space program and have developed in all educational levels. High schools and universities are beefing up their science curricula for both teachers and students with the assistance of NASA, and professional societies such as The National Space Society, The Planetary Society and the American Institute of Aeronautics and Astronautics.

There are also a number of special programs in which young people can participate, such as model rocketry competitions and "astronaut training" programs for various age groups.

Our personal serendipities include our long life together and our sons and now our grandchildren.

Mary Jane and Randall Chambers as they celebrated their 50th wedding anniversary. Our long and happy marriage is a personal serendipity and we are still trying to answer the question: which is harder – being married for 50 years or helping to figure out how to get off the planet.

Our family has been an example of the struggle of many modern families—balancing the tremendous day-to-day needs of family life with the demanding responsibilities of a challenging career—sometimes two challenging careers.

It is true that these days our homes are filled with many labor-saving devices unheard of in previous generations. But there still is no substitute for the loving presence of Mom and Dad. The challenges of raising a family seem to be ever-increasing. Rearing two fine young men is a major accomplishment of lasting importance.

It is often difficult for those born after 1950 to even imagine the state of science and technology at that time. Space research has brought about innovation and improvements in uncountable ways. Now in the 21st century new generations of scientists are continuing the legacy bequeathed to them by the space pioneers. Advances in science and technology— hard won by research and testing for the first astronauts—are being continued by eager young researchers.

Much has been written by space enthusiasts who are frustrated with the slow progress of the space program. At times it appears that our space explorations may not progress as we had envisioned.

What seems to be overlooked is the enormous benefits—the spin-offs—that space research has contributed to improving life and health on this planet.

In manned space research it may be the journey—the quest for knowledge—and not the destination which is the spectacular achievement.

The space journey also includes the truly awesome accomplishment of returning safely to Earth—our home! We have just begun our reach for the stars.

Our sons, Craig, left, and Mark, are all grown up now with lives of their own. They are still bringing countless joys to our lives.

Our oldest grandson, Randall J. Chambers, demonstrated his genes for space science at an early age. He is shown here in 2002 launching his "Honest John" rocket in a competition after studying physics and aerospace science as a hobby. (Photo by Thomas Beach).

Mary Jane and Randall growing old together on this planet, with enthusiastic support for space research, training and flight.

Those of us who were at the right place at the right time to begin the quest would not exchange the experience for anything.

To the new generations of scientists, engineers, technologists, astronauts and other space fans we say, "Forward, onward, upward!"

May your knowledge ever increase. May your courage be unfailing. May your enthusiasm be boundless.

And your dedication to space missions be inspiring and unwavering.

Bon Voyage!